INTUITIVE IC ELECTRONICS

INTUITIVE IC ELECTRONICS

A Sophisticated Primer for Engineers and Technicians

Thomas M. Frederiksen, B.S.E.E.

Monolithic Data Acquisition Products
National Semiconductor Corporation
Santa Clara, California

McGraw-Hill Book Company

*New York   St. Louis   San Francisco   Auckland
Bogotá   Hamburg   Johannesburg   London   Madrid
Mexico   Montreal   New Delhi   Panama   Paris
São Paulo   Singapore   Sydney   Tokyo   Toronto*

Library of Congress Cataloging in Publication Data

Frederiksen, Thomas M
    Intuitive IC electronics.

    Includes index.
    1. Semiconductors.   2. Integrated circuits.   I. Title.
TK7871.85.F74   621.381′73   80–27910

ISBN 0-07-021923-0

    4567890   KPKP   89876543

The editors for this book were Barry Richman and Susan Thomas, the designer was Mary Ann Felice, and the production supervisor was Paul A. Malchow. It was set in Baskerville by The Kingsport Press.

Printed and bound by The Kingsport Press.

This book is dedicated to the field application engineers of National Semiconductor, who provided the initial stimulation and appreciation which was needed to make this effort seem worthwhile.

Contents

vii

List of Illustrations

Foreword

Today's design engineer is faced with a myriad of complex and expanding integrated circuit technologies. Trade journals are full of new devices and manufacturing techniques. The understanding of these techniques is usually associated with rigorous mathematical analysis and generally obscure semiconductor physics. Many authors have tried to provide an alternative view of why and how these devices perform as they do.

This book achieves that goal. Conventional terms and definitions are explained in a very graphic and visual way. This new perspective results in physical models that are easily related to their respective circuit elements. Through the deductive reasoning process, the models are then combined to form building blocks. These building blocks are combined into more complex structures and ultimately into the familiar technologies.

The author's insight into semiconductor phenomena is attested to by his many patents and his development of operational amplifiers that are standard for the industry. It is fitting that his enthusiasm and clarity of thought be transformed into a book of this type.

Roger W. Biros, Manager
Field Applications Engineering
National Semiconductor Corp.

Preface

The central idea of this book is to explain the basic semiconductor mechanisms that take place within solid-state integrated circuits in an easy-to-understand, intuitive way. This allows the reader to grasp the main ideas behind all the devices used in the new semiconductor products, from the many types of diodes to the bipolar transistors, the JFETs, the MOSFETS, and the developments in technologies for microprocessors and semiconductor memories. Semiconductor technology continues to move very fast; therefore, many new concepts and devices also have been briefly covered to provide a quick updating for the reader.

The explanations presented will aid both students, as a supplement to their texts, and many other people, who will find that this book assists them in attaining their desired level of understanding. The essentials of this book were presented by the author to a group of field-application engineers for semiconductor products. These people had worked with the application of ICs, and they were interested in learning more about the details of the inner workings. They were surprised to find that they could actually obtain a useful, intuitive understanding of the basic operation of a transistor from this short book. Far too many people, both inside and outside the electronics industry, find that they are using semiconductor products in their everyday lives and yet lack a basic understanding of how these products operate. This book will make these products understandable and, therefore, more "friendly."

To develop an intuitive understanding of semiconductors, we first establish a background for the solid state of matter in Chapter 1. For readers with a little knowledge in chemistry and the physics of solids, this will be a refresher in concepts already understood. We have extended this material to introduce the notions of an electric field and electric current. Crystals are introduced, and impurity doping of silicon is covered.

Chapter 2 builds on this physical framework and introduces the PN diode. All aspects of diode operation are covered, and many commercially significant diode mechanisms are explained.

Chapter 3 provides insight into bipolar transistor operation and discusses both small-signal characteristics and large-signal limitations. Some new ways to visualize and understand both the common-emitter connection (with both voltage and current drive) and the common-base connection are developed. These are used to provide an understanding of the output impedance for each of these connections. A simple model for the transistor is introduced and an example shows how this can predict the voltage gain of a single-stage amplifier.

Chapter 4 provides a short history of the evolution of transistor manufacturing techniques.

Chapter 5 shows how integrated circuits are built and introduces linear IC processes and both NPN and special PNP devices. The evolution of digital IC logic circuits is also highlighted, and the popular logic families are explained.

Chapter 6 provides insight into the junction field-effect transistor world. The basic operation of the JFET is covered along with the Bi-FETs,®* the new monolithic operational amplifiers. The high-frequency MESFET is also described in this chapter.

Chapter 7, the largest of the chapters, is devoted to the present workhorse of the semiconductor industry, the metal-oxide semiconductors, the MOSFETs. The advances being made by semiconductor memory and microprocessor chips are shown to be a result of an evolution from the earlier PMOS products. A new intuitive introduction to the operation of the MOS field-effect transistor is provided that allows an easy understanding of this complex device. The CMOS, DMOS, VMOS, and new N-channel technologies are all discussed, and logic circuits are described.

The essentials of charge-transfer devices are also presented: the bucket-brigade devices, the charge-coupled devices, and the charge-coupled imagers.

Semiconductor memory is highlighted, and the simple circuit that was the basis for the revolution in dynamic RAM is covered. Nonvolatile semiconductor memory products are also described.

* Bi-FET is a registered trademark of the National Semiconductor Corporation.

Acknowledgments

The author acknowledges the help of many colleagues over the many years during which most of these notions were formulated. The idea for this book originated a number of years ago as an outgrowth of an after-hours discussion with Jim Moyer. Later the idea was formalized when Roger Biros asked the author to provide an all-day tutorial presentation to the field-application engineers of National Semiconductor. The suggestion for company involvement came from Art Zias, and the immediate interest and enthusiastic reception of Roy Thiels, as well as the approval by Pierre Lamond, have very much helped this book live through the rough early days of its existence.

The special efforts of Barry Small for both constructive criticism of the early manuscript and a second critique of the final manuscript have been most helpful. The final manuscript also has benefited greatly from the careful reading and comments of Bob Pease. The assistance of the engineering staff of National Semiconductor, Tim Isbell, Wadie Khadder, Keith Russell, Dennis Morris, and Fred Smith, is also acknowledged. Other reviewers also assisted in helping to obtain a more understandable presentation and provided useful inputs that improved the text. These include Bob Dort, Andy Wolff, Vicki Frederiksen, Tim Skovmand, Judd Murkland, Bill Jett, Jim Congdon, Fred Jaccard, Julie Struss, and Jim Diller, Jr.

INTUITIVE IC ELECTRONICS

CHAPTER I

Physical Background Information

This chapter provides a basic framework for both the essentials of electricity and the physics of the solid state of matter. Most of the discussions will be about silicon, because it is the most common semiconductor material.

1.1 SOME BASIC ELECTRICAL CONCEPTS

Most people deal with electricity and electric appliances in everyday life, and these first-hand experiences can provide more basic information than is realized. We will develop the necessary background material for this book by making use of everyday examples wherever possible. By explaining these, it is hoped that a familiarity with electrical phenomena will result.

Electric current is basic to semiconductor devices. We will start, therefore, with a review of this important concept.

Electric Current

An *electric current* is the dynamic component of an electric circuit. Voltage sources create the "pressure" that causes current to flow. The things that are moving around in an activated circuit are *electrons,* which are very small negatively charged particles. Although an electric current is the flow of electrons, by convention, *current flow* is considered *positive* when it is in a direction that is *opposite to the flow of electrons,* as shown in Figure 1-1.

All material things contain electrons. Cathode rays, which "paint" the picture in a TV set, are streams of electrons that exist inside vacuum tubes.

1

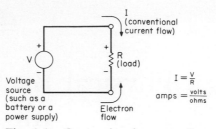

Fig. 1-1. Conventional current flow.

Electric Fields

The voltage applied to a circuit when a switch is closed establishes an electric field within a conducting material. In a toaster, for example, the electric field accelerates any available electrons in the heating element, thereby causing them to move. This movement of the negatively charged electrons constitutes an electric current. As the high-energy electrons collide with each other or with the material of the heating element, they lose some of their energy of motion (*kinetic energy* obtained from the electric field), transferring it to the heating element, which becomes *red hot* (600°C).

In the case of an incandescent light bulb, the filament becomes *white hot* (2000°C) and luminesces, or radiates, so strongly that we can read by the light produced. Inert gas within the bulb protects the filament, which would otherwise burn up (oxidize) at the extreme temperature. (A light bulb with a broken glass rapidly burns up if voltage is applied.) It is sometimes hard to believe that the light from a light bulb is produced by the collisions of conducting electrons within the bulb's filament.

Electric fields may seem to be a new concept, but you have seen their effects during lightning storms. When the electric field between a cloud and the earth becomes too intense, the normally insulating air that separates them breaks down (ioni*z*es) and allows current to flow. A large spark discharge called *lightning* results. Similarly, when you walk on a carpet in dry weather, you accumulate electric charges (as you do when rubbing a cat's fur with a comb) that set up an electric field between you and a metal door handle, for example. As your hand nears the handle, the electric field finally becomes intense enough to cause a surprising little electric spark, which *discharges* you, or removes the charge accumulated in your walk across the carpet.

Electric fields are developed from the electric forces that exist between separated groupings of electric charge. The smallest quantity of charge is that carried by an electron, -1.602×10^{-19} coulombs (the *coulomb* is the unit of electric charge). A current flow in an electric circuit is measured as the flow of coulombs per unit time.

If electrons are ripped or pulled away from a material, a restoring force, called the *Coulomb force*, is set up in that material. This force operates in such a direction as to cause the lost electrons to be returned (in other words, the material is now positively charged as a result of having lost some electrons). The strength of the Coulomb force varies as the inverse square of the distance separating an electron from the now positively charged material.

A *static field* is said to exist in a given region when electric charge is present in that region. Physicists have used this concept to simplify their work. By calculating the intensity and direction of the static field within a particular region, a physicist can account for the effects of all the fixed charge in that region. The electric (or static) field so derived is used to represent the net force that would result from all the individual charges in the region acting on a single unit charge introduced into the region. Charts can then be made that indicate the strength (and direction) of the electric field throughout the region, much like the elevation contours on a topographic map. Any new charge introduced into the region will therefore experience a force as a result of the existing electric (or static) field.

The concept of electric field is similar to the concept of gravitational field used by physicists to account for the attractive force of the earth's mass, which causes all other masses on the earth to have weight (the force of gravitational attraction). When we compare the magnitudes of the forces resulting from (1) gravitational attraction of two separated electrons (which is due to their mass) and (2) electrical repulsion of the Coulomb force (which is due to the electrons' like charges), *we find that the Coulomb force is 4.18×10^{42} times stronger than the gravitational force.* This is independent of the distance between the two electrons, since both forces vary as the inverse square of the distance. So, for electrons, we can usually neglect the effects of gravity because the Coulomb force is so very much stronger.

From 1909 to 1913, Robert Andrews Millikan (1868–1953) used both the Coulomb force and the force of gravity to measure the charge of an electron. He thought of attaching an electron (or a few electrons) to a small oil drop more massive than the electrons (Figure 1-2). When the voltage V between the two parallel plates shown in the figure was adjusted until the gravitational force was balanced by the Coulomb force, the drop (as viewed with a horizontal microscope) hung motionless. This value of V was recorded. When the voltage was abruptly removed, the oil drop fell (as a result of gravity) and the mass of the drop was measured. The charge of the oil drop was determined by calculations

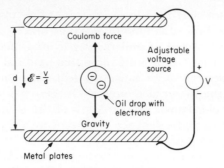

Fig. 1-2. Millikan's oil-drop experiment.

based on what strength of electric field *(V/d)* was needed to make the Coulomb force (directed upward) counterbalance the gravitational force (directed downward). In a series of experiments, Millikan found a whole-number relationship between the values of charge that were obtained. He deduced that he was dealing with multiples of a basic *unit* of charge, the charge of a single electron. In this way not only did Millikan show the discrete nature of the electron (it only exists as 1*e,* 2*e,* 4*e,* etc.) but also he very accurately measured the charge of a single electron.

Electric Power

The electric power supplied to a resistor depends on both the voltage across the resistor and the magnitude of the current that is flowing. It is only when a voltage source is allowed to cause a current to flow that electric power is consumed.

Energy must somehow be transformed to create a voltage source which will supply a current flow. Batteries use chemical energy to create voltage across their terminals. Voltage, therefore, is a form of electric potential energy, and some energy source is needed to create it. When this voltage is used to cause a current to flow (for example, in a flashlight bulb), electric power is being used and additional energy must be consumed if the original voltage magnitude is to be maintained.

Electric Energy

Electric energy is the product of electric power and time. It is therefore a way to measure both electric power and the length of time during which the electric power is consumed. The commercial unit of electric energy is the kilowatt-hour, and this is what utility bills are based on.

In science, the joule, or watt-second, is the unit of electric energy, and 1 kilowatt-hour is the same as 3.6×10^6 joules.

The unit of electric energy is named after James Prescott Joule (1818–1889), an English brewer. An amateur physicist in his spare time, Joule demonstrated by experiment that when the potential energy of a falling weight was used to turn paddle wheels in a water tank, it heated the water. This proved that heat was a form of energy rather than some magical fluid that flowed from one body to another, as was then commonly believed.

Resistance

It is obvious that a toaster and a lamp cause different power drains when plugged into the same 110-V wall outlet. The difference in power drain results from the different ability of each device to *resist* the flow of electricity. The name for this is *resistance,* measured in ohms (Ω). The *low resistance* of the toaster *causes large values of current I,* measured in amperes (A) and sometimes abbreviated as amps, to flow.

Similarly, if one touches a radio or a lamp while sitting in the bathtub, the low resistance of the wet skin can allow sufficient current to flow through the body to stop the proper functioning of the heart and result in electrocution. The current flows because of a low-resistance path from the 110-V wires through the wet body to a ground, which is supplied by the water pipes. Dry skin raises the resistance of the body and therefore reduces the current flow, but a 110-V source can still be fatal and should not be touched. Incidently, the danger is greatest where plumbing is exposed, such as in kitchens, bathrooms, or laundry areas. Conventional circuit breakers (which only limit the maximum current flow) are no protection.

Current Flow in Metals

Most metals have so many electrons available for conduction that they have small resistance values. Metal wires (usually copper, although silver is the best conductor of all the metals) are therefore used to connect a lamp to a power outlet. Because these wires have such low resistance values, they do not get nearly as hot as the filament in the lamp. Larger-diameter wires (with even lower resistance) are necessary for higher-power loads (such as a toaster) to minimize voltage loss at these higher currents.

To give you some familiarity with relative magnitudes, we can calculate the value of the electric field necessary to carry a current of 1 ampere in a copper wire 0.3 centimeters ($\approx\frac{1}{8}$ inch) in diameter and 5 centimeters ($\approx$2 inches) long. This is diagramed in Figure 1-3, and we start by calcu-

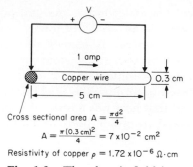

Fig. 1-3. The electric field in a short section of a wire.

lating the resistance R of this length of wire, which is given by

$$R = \frac{\rho l}{A}$$

where $\rho =$ the resistivity of copper (1.72×10^{-6} ohm—centimeters, a basic property of the material)
$l =$ the length of the wire
$A =$ the cross-sectional area of the wire
Substituting the values for this example, we have

$$R = \frac{(1.72 \times 10^{-6} \text{ ohm—centimeters}) (5 \text{ centimeters})}{7 \times 10^{-2} \text{ square centimeters}}$$
$$= 1.2 \times 10^{-4} \text{ ohms}$$

The voltage drop V across this R when carrying 1 ampere is given by Ohm's law:

$$V = IR$$
$$= (1 \text{ ampere}) (1.2 \times 10^{-4} \text{ ohms})$$
$$= 1.2 \times 10^{-4} \text{ volts}$$

This is a very small voltage. The electric field $\mathscr{E}$ within the copper wire (with a length l) can now be found as

$$\mathscr{E} = \frac{V}{l} = \frac{1.2 \times 10^{-4} \text{ volts}}{5 \text{ centimeters}} \times \frac{10^2 \text{ centimeters}}{\text{meter}}$$
$$= 2.4 \times 10^{-3} \text{ volts per meter}$$

which is also a very small electric-field strength. This example shows that it is difficult to establish high values of electric fields in conductors because of the resulting large values of current that would flow.

Another interesting calculation can be made to discover the average velocity of the conducting electrons in this same example. We start with the definition of an ampere of current flow:

$$1 \text{ ampere} = \frac{1 \text{ coulomb}}{1 \text{ second}}$$

A coulomb is a measure of electric charge, and the ampere is therefore a measure of the charge flow per second. The charge carriers in this example are electrons, and each electron carries the very small charge of -1.6×10^{-19} coulombs. We further need to know that most metals have approximately 10^{23} electrons per cubic centimeter of volume. Now we can find the average electron drift velocity v as

$$v = \frac{1 \text{ coulomb}}{1 \text{ second}} \times \frac{1 \text{ electron}}{1.6 \times 10^{-19} \text{ coulombs}} \times \frac{1 \text{ cubic centimeter}}{10^{23} \text{ electrons}} \times \frac{1}{A}$$

where A = cross-sectional area (0.07 square centimeters in this example).

So

$$v = \frac{10^{19}}{(1.6 \times 10^{23})(7 \times 10^{-2})} = 8.9 \times 10^{-4} \text{ centimeters per second}$$

or

$$v = 8.9 \times 10^{-4} \text{ centimeters per second} \times \frac{1 \text{ inch}}{2.54 \text{ centimeter}}$$

$$= 3.5 \times 10^{-4} \text{ inches per second}$$

As can be seen, there are so many electrons available for conduction in a metal that they do not have to move very rapidly in a given direction to represent a large current flow. This directed motion of the electrons, which is necessary to carry a current, is usually small when compared with the larger random motions of electrons resulting simply from their thermal energy.

Johnson Noise

While we are discussing the movement of electrons, it is worth mentioning that the *random motion of conducting electrons,* which, as mentioned, is due to their thermal energy, *gives rise to fluctuating signals or noise voltages* and forms the basis of *thermal,* or *Johnson, noise* in resistors. This noise, plus the additional noise of active devices such as transistors, limits the maximum sensitivity that can be obtained in electronic amplifiers. As expected, therefore, when listening for the small signals sent back to earth from the Venus-probe satellite, special measures must be taken in order for scientists to hear something other than background noise. Such measures include (1) using a narrow-band-width radio-frequency communication link, (2) using special low-noise receivers, and (3) greatly reducing the operating temperature of receiver components by cooling. You can see the effects of Johnson noise when you switch your TV to an unused channel. The only signal available is noise, which creates the random white spots on the screen and the hissing sound you hear coming from the speaker.

Electromigration

Conducting electrons also can cause some strange effects in certain metals. For large current densities, especially at elevated temperatures, *the metal will migrate.* Some atoms of virtually all metals are periodically broken free from their regular structure by thermal energy. If this occurs in a conducting metal, the "electron wind" of the conducting electrons can give up some kinetic energy (energy of motion) to the free atoms of the metal, causing the metal to migrate, that is, *physically move,* in the direction of the electron flow. Reliability problems resulting from migration can be created in integrated circuits if the current density in the on-chip aluminum interconnect leads is allowed to be too large. Increased cross-sectional area in the aluminum prevents this, and a glass-surface passivation layer on the aluminum prevents the more rapid movement that would otherwise result at the exposed surface. In addition, modifying the aluminum in the interconnect metal by introducing 4% copper also greatly reduces this migration problem.

Insulators and Semiconductors

Not all materials will conduct an electric current. *Insulators,* such as glass, rubber, and most plastics, are materials that have no electrons available for conduction and therefore conduct no current in the presence of an electric field. *Conductors,* such as copper or aluminum, have a large number of free electrons available for conduction. *Semiconductors,* such as germanium, silicon, and gallium arsenide (GaAs), lie between these extremes. They have some free electrons, but not enough to make them good conductors of electric current, especially when they are very pure.

It is the relatively strong bonding of the electrons in a silicon crystal that causes silicon to be classified as a semiconductor. When pure silicon is at room temperature, only approximately 1 out of every 10^6 of its atoms is ionized (which means that one of its electrons has acquired sufficient energy to break away from the atom to become a conducting electron). When silicon is heated, it becomes a better conductor because thermal energy "tears" bonded electrons away from the crystal lattice to produce charge carriers. Therefore, silicon is not a good insulator at high temperatures. The stronger electron bonding of the crystalline form of carbon, the diamond, causes this material to be a good insulator, even at relatively high temperatures.

Energy-Band Theory

Band gaps, allowed energy bands, and forbidden bands have to do with the amounts of energy electrons can have when they are in a solid. Electrons are strange things. If you isolate an electron in a vacuum, you can give it an arbitrary amount of energy. Many machines (electron accelerators) have been devised to do just this for experiments in nuclear research. The Stanford linear accelerator, for example, provides high-energy electrons with velocities that approximate the speed of light (energies of 10^{10} electronvolts, where the electron then starts acting much like a quantum of light, a photon, except that it has a much larger, and relativistically increasing, mass than the essentially massless photon).

If you are dealing with an electron that is associated with a single free atom, such as an isolated gas atom, then the electron can have only certain, definite amounts of energy. These are called *energy states,* and they correspond to the different possible orbits of electrons around the atom's nucleus. Each of the atom's electrons has a specific minimum-energy state, and only certain definite *higher-energy,* or *excited, states* are allowed.

A minimum amount of energy corresponding to the difference between two allowed states has to be supplied to the electron to *pump it up* to an excited state. Such "energetic" electrons are unstable and can *drop back* to other lower-energy states. When this occurs, the electron gives up its gained energy as electromagnetic radiation, which provide the characteristic spectra of the elements when examined in a spectroscope. Fraunhofer and Kirchhoff examined the spectra of sunlight and found that the dark lines in the spectra were due to the absorption of specific frequencies of light by vapor clouds that surround the sun. These same characteristic dark line spectra are the bright lines present in the emission spectra of certain elements. Each element has a characteristic "signature" or grouping of line spectra, and this has been useful in determining the elements present in stars, for example. Commercial use is also made of this in the colorful fireworks manufactured for Fourth of July celebrations.

When solids are formed, the discrete allowed energy states of their electrons appear to smear together to become *continuous bands* of energy. In actuality, however, the bands are collections of closely spaced energy levels. This smearing is caused by the proximity of adjacent atoms and the other orbiting electrons in the solid structure. (In fact, it is now known that electrons can even change positions from one atom to another, presenting new orbiting possibilities.)

The separations between energy bands in a newly formed solid structure and whether or not there are any electrons in the highest available band, the *conduction band,* determine whether a material is a conductor, an insulator, or a semiconductor. Conductors, such as metals, for example, are characterized by the fact that the highest-energy band that contains electrons, the conduction band, is only partially filled, as shown in Figure 1-4. When this is the case, it is easy to add essentially any

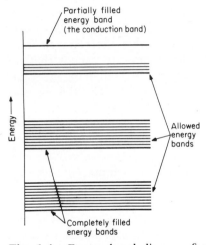

Fig. 1-4. Energy-band diagram for a conductor.

desired small amount of energy to the electrons in the partially filled band. The small added kinetic energy provided by an electric field is accepted by these electrons, and the directed motion which results causes an electric current to flow.

In the case of an insulator (a nonconductor, or dielectric), there is a wide separation (in energy) between the *last completely filled band* and the *next allowed, but empty, conduction band* (Figure 1-5). (Note from the figure that the energy gap is 9 electronvolts for the insulator silicon dioxide.) We cannot affect the energy of the electrons in the completely filled band because, since the band is filled, there is no way any of the electrons can accept additional energy. If we had *a lot of energy,* we could move electrons across the large band gap and obtain conduction. This is to say that insulators *will* conduct if a large enough electric field is placed across them, causing an electron to *jump the band gap*. Unfortunately, this is *dielectric breakdown* and usually results in permanent damage to a solid insulating material.

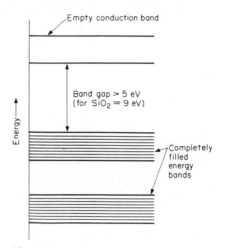

Fig. 1-5. Energy-band diagram for an insulator.

The smaller band-gap energy of the semiconductor silicon (1.1 electronvolts), for example, allows reasonable values of thermal energy (on peaks of the average thermal energy kT, where k is Boltzmann's constant and T is the absolute temperature in degrees Kelvin) to occasionally *kick* an electron across the 1.1-electronvolt band gap and thereby provide a current carrier. Semiconductors with larger band-gap energies, such as gallium arsenide (1.4 electronvolts), are correspondingly less likely to have free electrons. The large band gap therefore causes undoped (relatively pure) semiconductors to be poor conductors of electric current. As mentioned earlier, however, they are better conductors at higher temperatures.

The energy-band diagram for a semiconductor is shown in Figure 1-6. Because of the relatively poor intrinsic electrical conductivity, slight amounts of impurity doping (as we will soon see, where other elements replace a few of the atoms of the semiconductor material) cause dramatic effects in the conductivity of a semiconductor.

Let us now see how we can modify the semiconductor silicon for

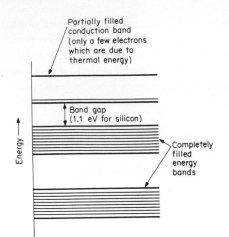

Fig. 1-6. Energy-band diagram for a semiconductor.

use in many electronic devices. To start, we will first take a look at some general characteristics of solids.

1.2 THE SOLID STATE OF MATTER

Chemists classify compounds and elements as existing in one of three (where, for simplicity, we omit "plasmas") possible physical states: gas, liquid, or solid. Most things exist in one of these forms, depending on the temperature of the material. A very familiar compound, water, for example, can exist in the gaseous phase as steam, the liquid phase as water, and the solid phase as ice. It usually happens that things progress through all three states from some cold temperature (the freezing or solidification point) to progressively higher temperatures. This orderly sequence need not be followed, even with water, as people who live in snow country realize. Clothes can be hung out to "dry" on a freezing day and water will go from the solid phase directly to the gaseous phase without passing through the liquid phase (*sublimation*).

What concerns us here are the interactions at the atomic level that cause these various states of matter to exist. It is easy to visualize gases as individual atoms or molecules (such as water vapor, H_2O) that are essentially not attached to each other but exist in the same physical container, much like a room that contains a few balloons, representing the individual elements or molecules, randomly moving about. The energy that supports the internal motion of the gas molecules is provided by some external heat source, and the collisions of the gas molecules with the walls of the container create the *pressure* of the gas.

Under reduced temperature, the gas will liquefy because the individual molecules or atoms will lose thermal energy and therefore tend to group together. Molecules have a basic attraction for each other at large distances which, interestingly enough, turns into repulsion at extremely short distances. It is this molecular repulsion that allows material to have strength and resist further compression.

At close spacings, the outer orbits of the electrons can almost overlap and cause an additional attractive force. This condition represents the liquid phase, where the molecules or atoms move easily over each other but are linked together.

For continued reductions in temperature, there is finally so little thermal energy present that the particles can finally assume a solid structure, owing to their attractive forces, and this is the *solid state*—the thing semiconductors are all about. Differences in the magnitudes of these attractive forces for various materials cause different physical states to exist at room temperature.

As the temperature is further reduced and approaches absolute zero ($-273°C$, or $0°K$) these attractive forces can ultimately hold any material *almost* rigidly together. [It is interesting to notice how *hot* and therefore *energetic* ordinary room temperature ($298°K$) is when compared with the terribly cold reference of absolute zero ($0°K$).] We say *almost*, because, even at absolute zero, the atoms still must have some motion to prevent violating the *uncertainty principle* of quantum physics, which states that *we cannot simultaneously know both the position and the momentum of an atomic particle.* Therefore, no matter how cold, the particles are still moving. (As a curiosity, helium remains a liquid at $0°K$ and ordinary pressures, and it has a boiling point of $-268.9°C$, or $4.1°K$.)

Having reached the solid state, we might look at the way an element such as silicon (Si), for example, goes about holding itself together with these interatomic bonds. This will require that we take a closer look at the atom and interatomic bonding.

An Atomistic View

The world of the atom is found to be very similar to our solar system. Both are made up mostly of empty space and have moving objects that are in orbits around a central object. For our solar system, the central object is the sun and the orbiting objects are the planets: Mercury, Venus, Earth, Mars, Jupiter, Saturn, Uranus, Neptune, and Pluto. Each planet has its own special path, and therefore, the planets travel only in certain layers, or orbits, around the sun. The force that holds this system together is called *gravitational force,* and it is the same force that keeps you on the surface of the Earth and causes you to weigh something on a scale. A scale measures the gravitational force (or attraction) the mass of the Earth has for the mass of your body.

Returning to the atom, we find that the central object is a positively charged *nucleus,* which contains essentially all the mass of the atom, and the objects that are in orbit are *electrons* (their appearance, more precisely, is "cloudlike"). The diameter of the nucleus is between 10^{-5} and 10^{-4} angstroms (Å, where 1 Å $= 10^{-10}$ m), and the much larger area occupied by the orbital electrons is illustrated by the fact that the nuclear diameter is only 0.001 to 0.01% of the entire atomic diameter.

The very small mass of the orbiting electrons causes their gravitational attraction to the nucleus to be negligible. However, since the nucleus is positively charged and the electrons are negatively charged, the forces that hold the atom together are electric and result from the differences in charge polarity, as shown in Figure 1-7. These same types

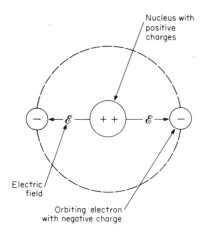

Fig. 1-7. A simplified diagram of the helium atom.

of electric forces also hold all material things together, because material is basically extremely large groupings of atoms.

These attractive forces are extremely strong in the single-element, or valence, crystals that form in the case of carbon and silicon. In the crystal form of carbon (the high-cost diamond), the greater strength of the forces within the crystal lattice allows the diamond to break up or scratch most other materials (*note how the selling price increases as we go from an amorphous form of carbon, soot, to a crystalline form, diamond*).

Crystals that have strong bonding typically are fragile and do not bend, even though they are sometimes very difficult to break. The different crystalline form of such metals as gold, copper, or silver results from weaker bonding and allows these materials to be easily bent or deformed. In general, only a *sea of electrons* holds metal atoms together; no precise atomic order or direction (crystalline lattice) is seen in their bonding. (This "lack of concern" among neighboring atoms is also why metals easily form alloys.)

Silicon or carbon do not always exist in a crystalline form. They can exist as amorphous carbon (graphite) and amorphous silicon, where the atoms are still bound by valence electrons, but without the precision or regularity of order that gives the long-range structure (the lattice) found in a crystal. However, each atom still evidences a localized order on a small scale. In addition, polycrystalline materials exist which are composed of aggregates or clumps of crystalline material.

The crystal form of silicon is diagramed in Figure 1-8, where the silicon atoms are represented by the internally labeled balls. The sharing of electrons with adjacent silicon atoms is what chemists call *covalent bonding*. In the figure, the solid line between each electron symbol and a silicon atom signifies which atom donated that electron. In actuality, however, we cannot put such tags on individual electrons; original ownership of electrons is both unknowable and of no consequence.

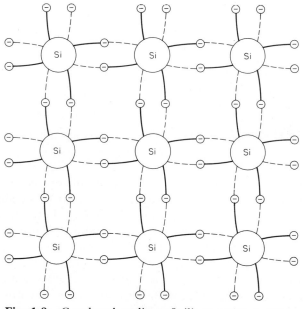

Fig. 1-8. Covalent bonding of silicon atoms to form a silicon crystal (a two-dimensional view).

Covalent bonding and the crystal state of silicon must be understood before an understanding of semiconductors is possible. To be accurate, we should mention that the bonding of silicon atoms continues in the third dimension (out of and into the page, for example) to provide a three-dimensional solid (with length, width, and depth), as shown in Figure 1-9. This is called a *diamond-crystal structure* after the crystalline form of the element carbon.

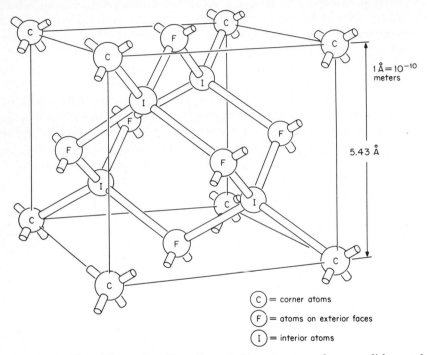

$1 \text{Å} = 10^{-10}$ meters

5.43 Å

(C) = corner atoms

(F) = atoms on exterior faces

(I) = interior atoms

Fig. 1-9. Three-dimensional bonding of silicon atoms to form a solid crystal.

In the covalent bonding of the element silicon, each atom shares its four outermost electrons, *valence electrons,* with other silicon atoms. These shared electrons are represented by the rods between neighboring silicon atoms in Figure 1-9. To form the crystal structure, two covalent bonds are involved between each pair of atoms.

Readers with a background in chemistry will recognize that this results in what is called the *stable condition,* where eight valence electrons are associated with each atom in the crystal lattice. The outermost shell (or band) of electrons (the valence electrons) controls the chemical activity of an element. The elements that naturally have eight valence electrons are therefore chemically inactive. This is the case for the "noble" elements, or inert gases, such as neon, argon, and krypton.

Heat Conduction

In metals, heat is conducted by the mobile outermost electrons. These *valence electrons* gain thermal energy and then transfer it to other electrons and to the lattice. This is why the thermal conductivity of metals is closely linked to their electrical conductivity.

Heat conduction also can occur by vibrational energy, which can be mechanically transferred from atom to atom through a crystal lattice. In insulators and semiconductors, these lattice vibrations conduct the heat. In such cases, heat conduction has nothing at all to do with the electrical conductivity of the material. For example, the tightly bound crystal of a diamond is a better thermal conductor than copper.

The Material Silicon

Silicon is a metallic-appearing solid that breaks easily and produces sharp edges at the break, similar to glass. It melts at 1412°C, and some high-temperature processing of silicon is done at temperatures in excess of 1200°C (in quartz tubes). At this elevated temperature, the silicon glows radiantly.

Silicon contains approximately 5×10^{22} atoms per cubic centimeter. To be useful in the manufacture of semiconductors, silicon must initially be extremely pure. Only one out of every 10^{10} silicon atoms can be an impurity. To appreciate the contamination problem in the manufacture of semiconductor products, this degree of purity is equivalent to 1 grain of sugar in 10 buckets of sand.

Next to oxygen, silicon is the most abundant element on earth. It makes up about 25% of the earth's crust and is found only in combination with other elements. Many familiar rocks are the oxide of silicon (SiO_2): sand, quartz, amethyst, agate, flint, jasper, and opal. Such materials as granite, hornblende, asbestos, feldspar, clay, and mica are only part of the list of rocks that contain silicon. When silicon combines with carbon to form silicon carbide, it then becomes a very tough abrasive.

Hyperpure silicon can be prepared by the thermal decomposition of ultrapure trichlorosilane gas ($SiHCl_3$) in a hydrogen atmosphere. Float-zone purification processes are also used to reduce impurity content. In such processes, a localized molten region (the float zone) is moved across the solid sample and acts to "sweep" all the impurities down to one end, (because the recrystallization excludes the contaminants) where they remain in an unused part of the sample.

This relatively common element has become the main ingredient in modern-day semiconductor components. Silicon is used in a wide variety of products from small signal diodes to extremely complex, very large scale integration (VLSI) memory and microprocessor chips. We will now look a little closer at the effects of impurity doping on the electrical characteristics of single-crystal silicon.

1.3 IMPURITY DOPING

Here is where the action is! The relative uselessness of intrinsic (pure) silicon (as far as carrying an electric current is concerned) can be greatly changed by the substitution of some different atoms within the lattice. This is called *impurity doping*, and the modified silicon is called *extrinsic*, or *doped, silicon.*

Very dramatic changes take place in electric conduction as a result of even slight amounts of special impurity doping. These impurity dopants take the place of an extremely low percentage of silicon atoms within the silicon crystal. In relatively light impurity doping (concentrations of 10^{15} atoms per cubic centimeter), only 1 out of every 50 million silicon atoms is replaced in the crystal. In very heavy doping (concentrations of 10^{20} atoms per cubic centimeter), 0.2%, of the silicon atoms are substituted by an impurity dopant.

At room temperature, these impurity atoms are *ionized* (or excited) by thermal energy, making them available to conduct current when an electric field is applied. In silicon, both positive and negative charge carriers are possible. The particular dopant used determines the type of charge carriers that result.

Impurity Types

Two types of dopant are possible for substitution in the silicon crystal lattice. We could choose an atom with five valence electrons, one more than the four that silicon has. This would provide an extra electron that does not fit into the covalent bonding of the silicon lattice and therefore can be used to conduct current. The doping atoms *phosphorus* (P), *arsenic* (As), and *antimony* (Sb) are called *donors* because they possess this extra electron. When silicon contains donors it is called *N-type silicon.* The N denotes negative and is used to represent the surplus of negative charge carriers, the electrons, now available.

If we choose a doping atom with only three valence electrons, such as *boron* (B), a place exists in the lattice for a fourth electron; this dopant is therefore called an *acceptor*. When silicon contains acceptors, it is called *P-type silicon*. The P denotes positive and represents the surplus of "places for electrons" (called *holes*) that now exist. Acceptors therefore act as positive charge carriers.

A small section of the periodic chart of the elements is shown in Figure 1-10. This shows that carbon (C) is in group IV with germanium

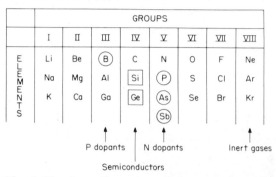

Fig. 1-10. A small section of the periodic table.

(Ge) and silicon (Si), which explains the similarity between these three elements. The N-type dopants come from group V (valence of 5), which means they have an extra electron when compared with an atom of silicon. Similarly, the P-type dopant boron (B) comes from group III. (Note that the inert gases are in group VIII; they have a completely filled valence shell of eight electrons and are therefore chemically inactive.)

For now, we will not concern ourselves with how doping is done in such extremely small concentrations. Let us look, rather, at the end results of impurity doping. First, remember that the distinction between the two basic doping types, valence of 5 or 3, is where the N's and P's come from in PN diodes and PNP or NPN transistors.

The concept of doping can be extended even further. For example, a piece of silicon can be initially doped with antimony (Sb) and therefore will be an N-type silicon. Then a P-type dopant, such as boron, can be selectively introduced into the N-type silicon. The selected areas of P-type doping will then *balance out* or *compensate* the previous N-type impurity doping. Increasing amounts of P-type dopant will then cause the silicon to be converted to P-type. Further, if a higher concentration of N-type phosphorus is then selectively introduced into the recently formed P-type regions, these will again be converted back to N-type. This sequence of subsequent type conversions, or *overdoping* (Figure

1-11), is basic to the manufacture of integrated circuits, transistors, and diodes.

Impurity-doped silicon with both N-type and P-type regions is also used for resistors in integrated circuits. The larger resistor values are made with lightly doped silicon because fewer charge carriers raise the resistivity.

When silicon is not doped (intrinsic silicon), thermal energy creates electron-hole pairs in equal numbers. This provides an intrinsic concen-

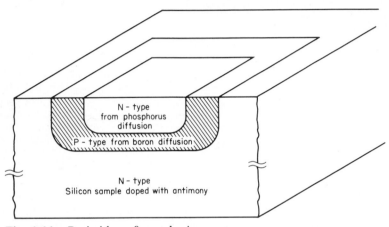

Fig. 1-11.   Basic idea of overdoping.

tration of charge carriers n_i, where n_i equals p_i. Under thermal-equilibrium conditions, the product of the hole and electron concentrations for doped silicon remains constrained to the square of this intrinsic, temperature-dependent value (n_i^2). For an N-doped piece of silicon, we find that the background concentration of holes will be depressed below the intrinsic equilibrium concentration because the added electrons from the doping material "fill" some of the holes. Therefore, even in doped silicon, both carrier types are still available, although one type can be very definitely in the majority.

1.4 DIFFUSION OF MATERIALS

When one material can easily be mixed into or combined with another material, it becomes possible for diffusion to occur between the materials. *Diffusion* is the process by which mixtures of easily flowing materials naturally reach a final *blend* where the resulting material tends to be uniform in composition. Two different gases or two different liquids generally diffuse into each other quite readily, and as we shall see, this can be the case for certain solids.

Diffusion processes involving high temperatures have important applications in industry. If high temperatures are required to get the diffusion to occur, then it is possible to *freeze* or stop the process at any desired intermediate state. The material would be an incompletely diffused mixture. This incomplete diffusion is essentially how impurity doping of silicon is controlled.

Gas-Phase Diffusion

Diffusion is not a new concept. If a person lights up a cigar in a closed room, the odor soon *diffuses* throughout the room. This is an example of a gas-phase diffusion. We can learn about diffusion in general by considering this simple example. First, it is obvious that particles of smoke diffuse from a place of high concentration (the cigar) to places of lower concentration (the corners of the room). In addition, it has been found that the rate at which the particles move is a function of the differences in concentrations between the two points divided by the distance between the two points; this is called the *concentration gradient*. For example, four cigar-smoking poker players around a small table would "get" to you in a remote corner of the room *much faster* than if there were only *one* smoker. In this example of gas-phase diffusion, the diffusion will continue, although at a lower and lower rate of flow, until complete equilibrium is achieved, that is, the whole room is uniformly contaminated. (For the sake of simplicity, we have neglected the complicating effects of convection, which cause the air near the ceiling to be contaminated sooner than the air near the floor.)

Solid-State Diffusion

The solid-state diffusion of one material into another, even in very small percentage concentrations, causes dramatic changes in the characteristics of the final new material. Steelmaking, for example, utilizes solid-state diffusion. The accidental discovery that something very "magical" happened to *iron* after high-temperature heating in the presence of certain *organic* materials was a major step forward for humanity. Today we know that *steel* is formed when *carbon is diffused into iron.* You can imagine how surprised the first soldier must have been when a medieval solid-state physicist stepped up with a new "patented steel sword" and *simply cut off the iron sword* the soldier was carrying!

In the beginning, the "black art" of steelmaking was very mystical;

many so-called experts developed special formulations, rituals, and recipes for its manufacture. Today we know that most of the magic of, say, early Sheffield steel had to do with the presence of local trace impurities in the raw materials used in the steelmaking process. Such trace elements as vanadium, chromium, and tungsten contributed to the custom-tailored *high-speed steels* that brought about the machine age. The new steels were stronger than plain *carbon steels* and allowed higher metal cutting rates for the metal machining industry. Even today, anyone who owns an electric drill knows the advantages of HS drill bits over the lower-cost, carbon-steel bits.

The semiconductor industry has undergone a similar evolution. Today, semiconductor products are manufactured by solid-state diffusion of impurity dopants. Extreme cleanliness and overall contamination control are required, as well as process control, to make such operations more precise and less mystical.

Now we will look at the effects of diffused impurity dopants on the electrical characteristics of silicon.

1.5 EFFECTS OF IMPURITY DOPING

Impurity doping of silicon is the process of introducing controlled amounts of specific atoms into the regular structure of single-crystal silicon. Only certain atoms will create the desired change in electrical conductivity. As mentioned earlier, one of the characteristics of semiconductor doping is that there are two different ways of increasing electrical conductivity: by adding either extra *electrons* or by creating extra *holes*.

N-Type Doping and the Electron

Returning to the doping of silicon by phosphorus (N-type), we find that phosphorus has a valence of 5, which gives this atom one extra electron when it takes the place of a silicon atom in the silicon lattice (Figure 1-12). Four of phosphorus's valence electrons participate in the covalent bonding with adjacent silicon atoms in the crystal. Since the fifth electron is not tightly bound within the lattice, it can be easily removed from the vicinity of its donor atom. The thermal energy needed to accomplish this *ionization* exists at room temperature. Therefore, the N-type silicon specimen has some electrons (negative charge carriers) readily available to conduct an electric current if wires are attached and an external voltage is applied. As expected, if the silicon is only relatively lightly doped (denoted by N⁻), it is not as good a conductor of current as if it were heavily doped (denoted by N⁺).

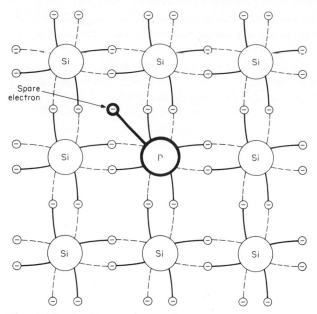

Fig. 1-12. An N-type dopant has a spare electron.

N-type silicon conducts an electric current in the same way a wire does: both have conducting electrons. The major difference is that there are not nearly as many electrons available in the silicon sample, so the resistance (and the resistivity) of the silicon is higher.

When an electron leaves the vicinity of a contributing phosphorus atom (which is stuck in the silicon crystal lattice), the atom has a positive charge because the original positive charge of 5 on the atom's nucleus (which is usually balanced by the presence of all five valence electrons) is now unbalanced. The phosphorus atom therefore has a *fixed,* or *immobile, net charge* of +1 as it sits locked in the silicon crystal lattice.

Under normal conditions, this charge imbalance does not really exist. An externally applied electric field (produced, say, by a battery or a power supply) causes electrons to move through the crystal, so at any given time an electron is near enough to each isolated phosphorus atom to balance the charges. In other words, the fixed positive charge is balanced by a mobile negative charge. The existence of both *fixed* and *mobile* charges is important to remember because it will be needed for an understanding of the action of PN diodes.

Confusion often surrounds the actions of these two types of charges. Only the *mobile charge carriers* are involved in *current flow.* The polarity of the fixed charge is always *opposite* to that of the associated mobile charge carrier. Large *electric fields* must be established within the silicon material to completely remove the *mobile charges* from the *fixed charges*. When large fields are used, however, *two basic types of regions* become possible in doped silicon: (1) *conductive regions,* where mobile carriers exist, and (2) *dielectric regions,* where only fixed charges exist. These properties make doped silicon useful not only for *resistors* and *capacitors,* but also for *amplifing* and *rectifying* devices.

To add to the confusion, two types of current flow are possible. One is similar to the current flow in a wire. This is a *drift current,* which flows as a result of an electric field. The other type of current flow is a *diffusion current,* which flows as a result of *a concentration gradient of charge carriers* (we will consider this later when we discuss diodes and transistors).

An example of electrons diffusing (mutually repelling and therefore spreading away from each other) is given in Figure 1-13. Notice that

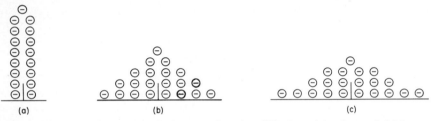

Fig. 1-13. Two-dimensional electron flow by diffusion: (*a*) a large initial gradient; (*b*) a short time later; (*c*) after more time has elapsed.

if we could somehow set up an initial "stack" of electrons in a small region (Figure 1-13*a*), as time passes, the electrons will move as a result of the concentration gradient (constituting a *diffusion-current flow*). This is similar to the solid-state diffusion of impurity dopants. In the doping process, impurities move from a high-concentration region, (around the silicon wafer in the diffusion furnace) to a low-concentration region (inside the silicon crystal structure).

P-Type Doping and the "Hole"

Things are not quite as straightforward when we consider a P-type dopant, such as boron. Boron has only three valence electrons; therefore, one of the neighboring silicon atoms comes up short when a boron atom replaces a silicon atom in the crystal. This absence, or *deficiency,* of an electron is called a *hole* because there now exists a *place for a spare electron,* if such a spare can be found wandering around in the silicon crystal.

While, in fact, this *hole* is the absence of an electron, it can be more easily understood, conceptually, by considering it to be simply the equivalent of a positive charge that is free to move through the crystal. The problem of just what the hole really is and how best to think of it and its motion through the silicon crystal can be the basis of much argument and confusion in the minds of students. Suffice it to say that thinking of the hole as a mobile positive charge is the easiest way to handle the problem. The confusion results from the fact that holes move because adjoining atoms provide recombining electrons. These electron losses at each location cause "new" holes to appear.

This whole process has been likened to the way a bubble moves up through a liquid. You see the bubble move, but what you actually see is the absence of fluid. What is really happening, therefore, is that *the fluid is moving down as the bubble moves up.* So it is with the motion of holes: the neighboring electrons are moving in one direction, and this produces the motion of the hole in the opposite direction.

It is generally assumed that all metals carry current by making use of conducting electrons. Actually, however, it has been found that the metal berylium conducts current by the action of *holes, not* by electron motion.

Majority and Minority Charge Carriers

The positive charge carrier, the hole, does not usually exist in metals, but it is the basis for the usefulness, and strangeness, of semiconductors. Because of impurity doping, we now have two charge carriers to work with, electrons and holes. Furthermore, with two charge carriers, we therefore have *majority* and *minority carriers.* For example, *electrons* are *majority* carriers in N-type silicon because N-type doping produces "extra" electrons. When electrons exist in a region of P-type silicon, they are called *minority carriers* because they are usually outnumbered by the higher concentration of holes produced by the P-type doping. Ordinary transistors are called *bipolar* because both minority and majority charge carriers are involved in their operation.

Minority carriers are the key to semiconductors. We will show how NPN transistors rely on them, as well as how they help explain many of the strange happenings in field-effect transistors. Much confusion exists about minority and majority carriers, and many people feel that discussions about these strange entities are really not needed, especially by electronics people.

Engineers and technicians are frequently bored with basic discussions of crystal structure, atomic bonding, and charge carriers; they are impatient to *get on with it,* to see how transistors really operate. However, you cannot develop an intuitive understanding of semiconductors until you feel very comfortable with these concepts. After all, a transistor or diode is a small, specially processed piece of silicon. Before you can understand why currents enter and leave such a solid structure, you must first understand the solid state of matter.

CHAPTER II

The Semiconductor Diode: The Basis of the Bipolar Transistor

Finally we have arrived at a point where we can put all the foregoing theoretical material together to forge an understanding of how and why a silicon PN diode works. If we were to take a cube of N-type silicon and somehow attach it to a second cube of P-type silicon, what would happen? Well, in essence, we have created a *diode*. For now, simply consider that one face of each cube is perfectly flat, and therefore, when these two cubes are pushed together, they have excellent atomic contact with no air spaces (Figure 2-1).

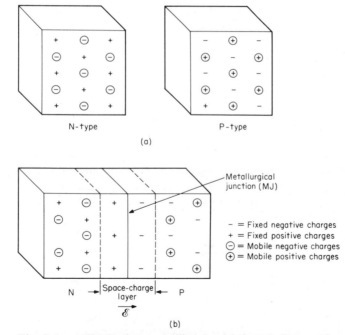

Fig. 2-1. A PN diode at equilibrium: (*a*) doped silicon cubes before contact; (*b*) cubes after contact and initial carrier flow. Note that the space-charge layer (SCL) has no mobile charge.

When the cubes are first placed in contact, the concentration gradients between the mobile electrons in the N-type cube and the holes in the P-type cube will cause both these types of charge carriers to diffuse across the contact surface. Before contact is made, each block obeys *charge neutrality;* that is, for each charge carrier bound in the lattice (fixed charge), there is a charge carrier of opposite sign free to move (mobile charge). The total numbers of each are the same in each block; an electric field is needed to unbalance this natural equilibrium condition.

After the blocks have been pushed together (Figure 2-1*b*), there is an initial short burst of mobile-carrier flow, by diffusion, across the boundary between the N and P sides (this boundary is called the *metallurgical junction, MJ*). Since there are more free electrons on the N-type side than on the P-type side, the *concentration gradient* on the N-type side causes the electrons to move (diffuse) into the P-type side. Similarly, the holes diffuse from the P-type side over to the N-type side. These mobile carriers are said to *recombine,* thereby mutually annihilating each other. Actually, the mobile electrons "fill" the holes, which, as you remember, are locations that can pick up a spare electron. This charge transfer across the metallurgical junction (MJ) unbalances the electric neutrality of the P and N regions; the regions near the MJ in both cubes no longer have any available mobile charge carriers. The fixed charge in this region of the lattice is not balanced; in fact, no available mobile carriers exist in this "depleted" region: the *space-charge layer* (SCL) (Figure 2-1*b*).

At first it might seem that this diffusion flow could go on forever: if we could attach a couple of wires, we could power a lamp in an external circuit—getting something for nothing, a perpetual-motion machine. Actually, here is what happens (and it is different from the simple gas-phase diffusion example). As the electrons diffuse away from the N-type cube, *they leave behind* positively charged immobile atoms (the nuclei of the donor atoms) stuck in the silicon lattice. Similarly, the diffusion of holes away from the P-type side leaves behind fixed negative charges embedded in the silicon lattice (actually, these negative charges exist as a result of the electrons trapped by the original holes).

The immobile charges *uncovered* in the lattice when this occurs set up a *built-in electric field that opposes any further diffusion* of mobile charge carriers of either type, as shown in Figure 2-1*b*. In Figure 2-1*a*, mobile electrons are shown as $\ominus$ on the N side and the mobile holes are shown as $\oplus$ on the P side. The charges that exist on the fixed dopant atoms in each silicon cube are shown without the circles.

The separated fixed charges exist in the same total number on both sides, and the electric field set up has the direction shown in Figure

2-1*b*. (The direction of an electric field is the direction in which an imaginary positive test charge would move if released in the region, that is, opposite to the direction an electron would move.) *The appearance of this electric field prevents any further large-scale diffusion of mobile charge carriers of either type.* Thus the tendency for the mobile carriers to diffuse across the junction is balanced by the built-in field resulting from the uncovered fixed charges. At equilibrium, therefore, this built-in field causes a drift-current flow that just balances the diffusion-current flow (the carriers are being returned at the same rate they leave). No net current crosses the junction. This is similar to water boiling in a covered pot. At thermal equilibrium, steam is being condensed on the sides of the pot and the cover, and water is thereby returned to the pot as fast as new steam is produced.

As we will soon see, applying an *external forward-biasing voltage* will *reduce* this built-in field, *allowing steady-state diffusion* of charge carriers across the junction (forward current). Similarly, an *externally applied reverse-biasing voltage* will *increase* the field across the junction (Figure 2-2), *preventing any net diffusion currents to flow.*

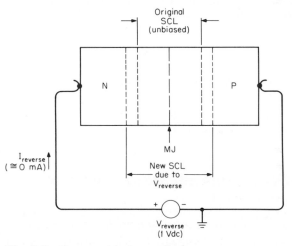

Fig. 2-2. Reverse-biasing a diode.

The appearance of this built-in field is a new phenomenon not mentioned in the gas-phase diffusion example. It is this phenomenon that keeps the PN junction diode from totally diffusing, completely eliminating the mobile carriers on the side where they are in smaller supply.

We can now consider the effects of nonequal impurity doping on each side of the diode. For example, we could have an N^+P^+ diode or an N⁻P⁻ diode, or other relative grades in between. In any case, since the separated charges across the space-charge layer (SCL) number the same on each side, a higher concentration of dopant on one side would cause a shorter penetration of the SCL into that side, as shown in Figure 2-3. If both sides are heavily doped, then a very narrow SCL results.

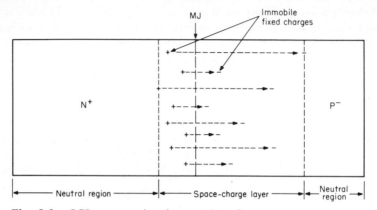

Fig. 2-3. SCL penetration in an N⁺P⁻ diode. Dotted lines with arrows show the direction of the electric field.

As we will soon see, heavily doped diodes have some interesting and useful characteristics. In addition, the impurity-doping levels on both sides of the diode determine which will be the dominant mobile charge carrier under forward bias, as well as the magnitudes of the breakdown voltage, junction capacitance, and reverse-biased leakage currents. These effects will now be considered.

2.1 WHY VARACTORS ARE MORE COMPLEX THAN A CAPACITOR

Varactor is an acronym for *variable reactor*. Any reverse-biased PN junction has a *depletion capacitance* associated with it, and this is where the *reactor* comes from: it is due to the capacitive reactance. Recall that a *capacitor* is simply two conductors separated by a dielectric. Since the SCL has no mobile charge carriers, it is the dielectric in this case.

The capacitance of the diode results from the additional fixed charge that widens the SCL under reverse-biasing conditions. It turns out, and we will shortly understand why, that the magnitude of the capacitance depends on many things: the doping levels on *both* sides, the details of the doping profiles (that is, the changes in doping density as one moves away from the junction), and the magnitude of the direct-current (dc) reverse-biasing voltage. The magnitude of the dc reverse-biasing voltage provides a way to control or vary the magnitude of the capacitance presented to an external circuit by the PN reverse-biased junction. This is where the *variable* comes from. As you can see, a *varactor* is a capacitor that can be varied or changed in magnitude by changing the dc reverse-biasing voltage.

This voltage-variable capacitance phenomenon is quite different from that of most commercially available capacitors. Actually, many disc-ceramic capacitors are also voltage-variable, but they do not exhibit the wide capacitance ratios attainable with varactors. For the simple parallel-plate capacitor with an air dielectric, the capacitance is fixed by the area of the plates, the separation of the plates, and the dielectric constant of air. The major difference with the junction capacitor results from the variable separations between the charges within the SCL. All the charges of the parallel-plate capacitor are separated by the same distance, the dielectric thickness. In the case of the junction capacitor, the charge is randomly embedded within the silicon crystal lattice, and essentially none of the opposing charge pairs within the SCL has the same separation. A comparison of charge separations in the two types of capacitors is shown in Figure 2-4. As can be seen, the junction capacitor does not result in as simple a geometric structure as the parallel-

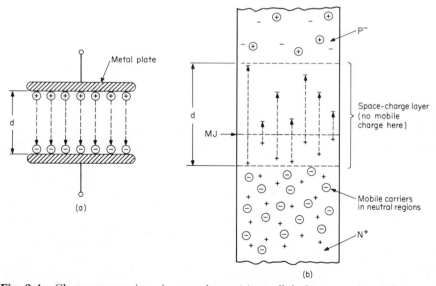

Fig. 2-4. Charge separations in capacitors: (*a*) parallel-plate capacitor; (*b*) junction capacitor.

plate capacitor. For example, as the reverse-biasing voltage is increased, the SCL in the junction capacitor widens because the externally applied voltage adds to the built-in field. Additional fixed charge is uncovered in the lattice *at the edges of the SCL.* The availability of this charge depends on the doping concentration. Usually the charges are provided at a non-linear rate as the reverse bias is increased; this creates changes in the capacitance of the PN junction that are a function of the inverse square root, or cube root, of the dc reverse-biasing voltage. Further, since the capacitance value is no longer a simple function of geometric dimensions, we must use Poisson's equation (which relates the charge to the electric field) to calculate the magnitude of the capacitance.

Now, to provide some feel for what is happening, we note that it requires a more intense electric field to increase the width of the SCL of a P^+N^+ diode (both sides heavily doped) because of the proximity of the opposite charges. For a P^-N^- diode (both sides lightly doped), these charges are more widely separated (a wider SCL); therefore a smaller change in electric-field strength will uncover additional fixed charges in the lattice. This indicates why the junction capacitance of heavily doped diodes is larger than that of lightly doped diodes for a given junction area. The simple parallel-plate capacitor is not as complex, since essentially all the charges are (always) separated by the same distance (the fixed separation of the plates), which does not change as the dc biasing voltage of the capacitor is changed.

In summary, notice that at low dc biasing voltages, the SCL is relatively narrow. The close spacing between the fixed charges on both sides of the junction therefore provides a large value of capacitance. The increasing separation of this charge at higher reverse-biasing voltages forms the basis for the decrease in capacitance.

Uses for Varactors

Varactor diodes are used for tuning high-frequency resonant circuits with a dc control voltage. They are also used as low-noise microwave amplifiers. Because the junction capacitor has only a small value of dc leakage-current flow through its terminals, *the shot noise* associated with dc biasing current in amplifying devices *is essentially not present.*

These "parametric" amplifiers are based on dynamically changing the value of a capacitor after a signal source has initially charged it. As an example of a parametric amplifier, an air-dielectric capacitor can be initially charged from a 1-V_{DC} signal source. The charge Q (in coulombs) that flows from the signal source onto the capacitor is given by

$$Q = CV$$

If the input signal is disconnected from the charged capacitor and the plates are mechanically pulled farther apart (by applying a force large enough to overcome the Coulomb force opposing this motion), the value of the capacitance is reduced. Now the same value of charge Q exists on a smaller-valued capacitor. The preceding equation indicates that for a constant value of Q, if C were reduced by one-half its former value, then V would double. This is what provides voltage gain.

The operation of a high-frequency parametric amplifier is based on having a *pumping frequency* dynamically change the reverse-bias voltage (and, therefore, the capacitance) of a varactor diode. The signal source (which is isolated from the pumping-signal source by special filter networks) charges the capacitance of the varactor when this capacitance value is relatively large. Then, because of further action of the also high-frequency pumping signal, the capacitance of the varactor decreases. Thus the initial charge provided by the input signal causes a larger signal voltage to appear across the reduced capacitance of the varactor. The pumping-signal source provides the energy necessary to reduce the capacitance. This results in a low-noise means of amplification because no noisy active devices (such as transistors) are used and no noisy resistors are involved.

2.2 LEAKAGE CURRENTS AND THE REVERSE-BIASED DIODE

Since the SCL is devoid of all mobile charge carriers, we would expect no dc current to flow under reverse-biasing conditions. In fact, a small current does flow. Since this current exists under the condition of an externally applied reverse-voltage bias (as opposed to a *forward-bias,* or *conducting,* condition), it is called a *reverse current,* or more popularly, a *leakage current.*

Why should a leakage current exist? There are *three basic reasons.* First, some unclean condition may exist (usually at the surface of the diode junction) that allows a conducting contaminant to shunt current across the surface of the junction. Currents of this type are typically small in value. In the early days of germanium transistors, low values of leakage currents were used to indicate "high quality" or long expected life of the transistor.

Second, in addition to the surface current (less of a problem in modern devices), existing *thermal energy* continually provides electron-hole pairs by occasionally breaking covalent bonds within the crystal lattice. For regions far removed from the SCL, these new electron-hole pairs recombine (mutually annihilate each other), having little effect on an external circuit.

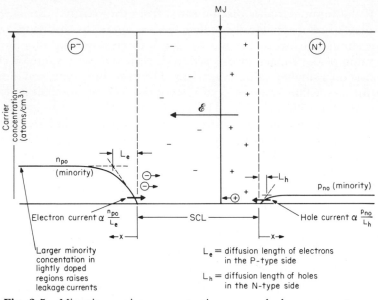

Fig. 2-5.   Minority-carrier concentrations cause leakage currents.

For those electron-hole pairs *generated near the SCL,* a different effect occurs, as shown in Figure 2-5. The majority-carrier concentration at thermal equilibrium on the P side p_{po} is smaller than the majority-carrier concentration at thermal equilibrium on the N^+ side n_{no} because this is a P^-N^+ diode. (The concept of thermal equilibrium implies that no external voltages are applied and that the material is in a settled steady state at the same temperature T as its environment.) An equilibrium background minority-carrier concentration is shown for each side of the diode as n_{po} and p_{no}, respectively. As previously stated, the p_{po}, n_{po} product equals the n_{no}, p_{no} product (both equal $n_i{}^2$) at any given operating temperature. Heavy doping, such as on the N^+ side in this example, therefore reduces the minority-carrier concentration on that side ($p_{no} < n_{po}$).

The diffusion lengths L_e and L_h are also shown in this figure. These are mathematical constants that indicate the shape of the exponential increase in the minority-carrier density as the distance x (shown on the figure) increases away from the edges of the SCL. The initial slope of these minority concentration gradients depends on the value of the diffusion lengths; smaller values of diffusion length provide steeper slopes.

Thermally generated carriers created within a few diffusion lengths of the edges of the SCL diffuse toward the SCL because the minority-carrier concentration is always zero at the edges of the SCL. This occurs because the relatively large electric field continuously existing in the

SCL "sweeps up" the minority carriers. However, minority carriers are always diffusing *toward* the SCL on both sides of the junction, and these constitute the second type of diode leakage current. In fact, at junction temperatures in excess of 100°C, this is the most significant type of leakage.

Finally, leakage currents are also *thermally generated within the SCL*. These currents are dominant at junction temperatures below 100°C. Since the width of the SCL increases as the reverse-biasing voltage increases, these currents are voltage-dependent. In general, leakage currents are smaller for more heavily doped PN junctions (if tunneling currents, which we will soon discuss, are avoided).

Temperature Dependence of Leakage Current

Increases in ambient temperature are exponentially related to increases in the thermal generation of electron-hole pairs. This forms the basis for the following rule of thumb: *Leakage currents approximately double for every 10°C increase in junction temperature.* This exponential dependence on temperature is likewise found with chemical reaction rates, which also double for each 10°C increase in temperature.

Nonideal Leakage Currents

There are many additional complications, especially in silicon, that contribute to or modify leakage currents. Fortunately, leakage currents in silicon devices are typically very low in magnitude. For our purposes, the typical doubling every 10°C is a sufficiently accurate first-order model to be useful for all but the most exacting cases.

It is interesting that most currents resulting from other causes do not exhibit strong exponential changes with temperature. Therefore, measurements of the change in leakage currents as a function of temperature can help isolate the internal causes of excessive leakage currents. This is especially true for collector-to-emitter leakage in transistor structures, which results from submicroscopic diffusions of the emitter dopant completely through the base width into the collector region. These diffusions are called *emitter pipes* and are the *Achilles' heel* of bipolar products. We will discuss emitter pipes in a later section, after the NPN transistor has been introduced.

2.3 REVERSE-VOLTAGE LIMITATIONS

Any insulating material eventually has problems holding off higher and higher applied voltages. This is why capacitors have maximum voltage specifications. The dielectric, or insulating, layer between the plates of a capacitor conducts no current simply because there are no free charge carriers available. The carriers are all tightly bound within the dielectric material. The problem is that as increasing voltages are applied, the electric field eventually becomes large enough to rupture the material and cause an electrical breakdown (as in the zener diode, which we will discuss in this section).

Electric-field strengths are measured in units of volts per meter of separation. Therefore, large electric-field strengths can be obtained with relatively low voltages if the separation between the conductors (the dielectric thickness) is sufficiently small. For example, a voltage of only 1 V_{DC} can create an electric-field strength of 10,000 volts per meter (V/m) if the separation is only 0.0001 m, because

$$\mathcal{E} = \frac{V}{d}$$

or

$$\mathcal{E} = \frac{1 \text{ V}}{0.0001 \text{ m}} = 10,000 \text{ V/m}$$

In the physical world we live in, we are more likely to think in terms of separations of ⅛ to ¹⁄₁₆ inch rather than in terms of the extremely small separations that exist within semiconductors, which are a few micrometers (μm), where 1 μm = 0.000001 m. Therefore, it is hard for many people to appreciate that the relatively low operating voltages of semiconductors (as compared with vacuum tubes, for example) can cause high-field breakdowns within semiconductor devices.

A PN diode is usually not in breakdown when no external reverse-biasing voltage is applied (as we will see later, the tunnel diode is the exception to this statement). As the external voltage is increased, the width of the SCL across the diode also increases. If both the external voltage and the SCL increase at the same rate, the electric field within the SCL stays constant. For example, if doubling the reverse voltage also doubles the width of the SCL, the ratio of these two, which is the electric field, stays the same. In lightly doped silicon, the spread of the SCL is greater than in heavily doped silicon, because the SCL has to pick up fixed charge, which is not as plentiful in lightly doped regions. The doping concentrations, therefore, affect the SCL spread. Thus a P^-N^- diode has a higher breakdown voltage than a P^+N^+ diode. Breakdown occurs when the electric-field strength builds up to a critical value within the SCL region.

Diodes in Breakdown

Special diodes have been designed and specified to operate continuously in a voltage-breakdown mode (Figure 2-6). These diodes are used as

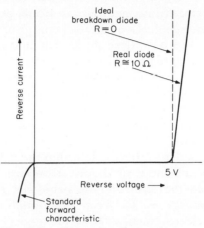

Fig. 2-6. Constant-voltage characteristics of a "breakdown diode."

voltage regulators (Figure 2-7) and are *all* erroneously called *zener diodes.* If we look at mechanisms responsible for electrical breakdown in silicon, we find that there are two: *avalanche* and *zener* (in honor of the American physicist, Clarence Melvin Zener, born in 1905). The strange thing is that these breakdowns occur at different values of electric-field strength.

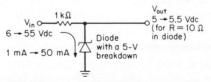

Fig. 2-7. A simple shunt voltage regulator.

Moreover, conditions within particular diodes usually favor one or the other, but both are simultaneously possible.

Avalanche Breakdown

Some readers (or their fathers) may remember the special voltage-regulator diodes, such as the OD3, VR-150, or even NE-2, that were the "zener diodes" of the vacuum-tube days. High-field breakdown within gas-filled tubes was first studied by Sir John Sealy Edward Townsend in 1901. He found this to be avalanche breakdown. The mechanism of avalanche breakdown is this: at a critical externally applied voltage, the electric field within the tube is finally intense enough that ionized gas atoms and free electrons (which are both generated by thermal energy) accelerated by this field smash into neutral gas atoms generating additional charge carriers. Random peaks of thermal energy are more intense than the average value, and it is these peaks that first ionize a neutral gas atom and get the whole regenerative cycle started. This is an *avalanching phenomenon,* and the end result is that at this critical externally applied (avalanching) voltage, the gas-filled diodes conduct relatively large currents. (Avalanching is not easily obtained in a conductor because extremely large currents would flow at the high value of avalanching electric-field strength.)

A mechanism similar to this exists within the solid material of semiconductors. In a semiconductor diode, an intense electric field exists across the SCL and the electron-hole pairs that result from random peaks in thermal energy are accelerated (although each in an opposite direction) by the electric field. The accelerated electrons then smash into the lattice and knock loose additional electron-hole pairs, resulting in a regenerative, or avalanching, breakdown.

Avalanching limits the maximum field within silicon (called the *critical field*) to approximately 3×10^7 V/m, or 30 V/μm. Therefore, an SCL width of 2 μm would support approximately 60 V. Values of avalanche breakdown depend on doping concentrations of the silicon, because higher doping reduces the *mean free path* (the distance between collisions) of the accelerated charge carriers. Avalanche breakdown values range from 30 V/μm (for doping levels of 10^{15} to 10^{16} dopants/cm^3) to 100 V/μm (as the doping concentration increases to 10^{18} dopants/cm^3). SCL widths for avalanche diodes used as voltage references vary from less than 1 μm to approximately 10 μm.

When an electric field is large enough to create avalanche breakdown, the velocity of the conducting electrons reaches a maximum or *limiting* value. Remember, it is the collisions of these conducting electrons with the lattice that cause resistors to become hot when current is flowing. A single lattice collision does not usually remove all the kinetic energy of an electron. Therefore, at relatively small values of electric field strength, the conducting electrons can easily pass through the lattice. But the velocity of conducting electrons steadily increases as the electric field strength is increased, and as the electrons gain more energy, a new energy exchange with the lattice finally becomes possible: *photon interaction.* The conducting electrons now can kick an electron of the lattice into a higher energy state, and when this excited electron subsequently drops back, electromagnetic energy (a photon) is emitted.

This new interaction reduces the energy of the conducting electrons to nearly zero, and to continue their passage through the lattice, they have to be accelerated again by the electric field. A maximum or *scatter limited* velocity v_s (approximately 10^5 m/s) is ultimately reached because this total energy exchange makes the lattice appear very dense since the velocity of the conducting electrons is repeatedly returned to zero. Carrier flow through SCLs usually takes place at v_s, and this highspeed carrier transport mechanism is made use of in many microwave applications of PN diodes. The impact-avalanche and transit-time (IMPATT) diode and the trapped-plasma avalanche and triggered-transit (TRA-PATT) diode are specially fabricated PN junctions that can provide gain at microwave frequencies when they are driven into avalanche breakdown by the signal voltage.

To appreciate why microwave frequencies can be attained, we can compare the magnitude of v_s with the magnitude of the drift velocity of the electrons in a wire from our example in Chapter 1 (8.9×10^{-4} cm/s). We find that v_s is larger by a factor of 100 million to 1. Electrons traveling at this extremely high velocity can cross an SCL which is 200 Å wide in much less than 1 pico-second (10^{-12} seconds), which, for example, is the period of a frequency of 1 THz (10^{12} cycles per second).

This avalanche form of breakdown, although incorrectly called *zener breakdown* most of the time, is the actual physical mechanism of breakdown if the magnitude of the breakdown voltage is equal to or larger than 5 or 6 V. Therefore, a *10-V zener diode* is not a *zener diode* at all, but more properly a *10-V avalanche diode*. "Zener diodes" are commercially available from 2 to 200 V, and most of them are really avalanche diodes. We will consider the true zener breakdown mechanism shortly.

One of the properties of the avalanche-breakdown mechanism is that it is basically noisy. This results from a new mechanism: charge generation from random collisions with the lattice. The amount of noise voltage generated increases directly with the magnitude of the breakdown voltage. As a result, both solid-state and gaseous avalanche diodes have been used as electronic noise generators. This noise is not always desirable, but it is, unfortunately, associated with this type of breakdown mechanism.

Temperature Effects on Avalanche Breakdown.   As temperature increases, there is more random movement of the atoms within the crystal lattice. This raises the probability of lattice impact by any thermally generated electron-hole pairs accelerated by the field. This reduction in the *mean free path* reduces the time duration within which the free carriers can be accelerated by the field. As a result, the carriers can only attain low velocities. On impact with the lattice, therefore, *no* additional electron-hole pairs are generated. Thus, at elevated temperatures, the externally applied voltage must be *increased* in order to establish the avalanching breakdown. As a result, avalanche diodes have a positive temperature coefficient of breakdown voltage ($+TC$). As a rough rule of thumb, this $+TC$ increases with the magnitude of the breakdown voltage at a rate of approximately 1 mV/°C of $+TC$ per volt of avalanche breakdown. Commercial use is made of this property in temperature-compensated reference-diode products, where one or two forward-biased diodes (with a TC of approximately -2 mV/°C each) are added in series with particular avalanche diodes to provide nearly zero temperature-coefficient (0 TC) voltage references.

Curvature-Limited Avalanche Breakdown.   The theoretical avalanche-breakdown limit of a PN junction is usually calculated for a plane diode. This is a parallel or sandwich structure with a continuous P-type region on one side and an N-type region on the other, with no considerations of breakdown-voltage problems owing to edge effects (Figure 2-8). When

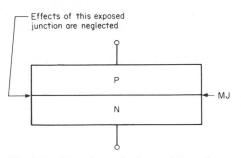

Fig. 2-8.   The plane diode used for voltage-breakdown calculations.

planar (diffused into one surface) diodes are fabricated, the perimeter of the diffusion curves up toward the surface, as shown in Figure 2-9. Note that deeper junctions (xj_2 in the figure) have less curvature at the edges because of their larger radius of curvature. Junction curvature reduces the breakdown voltage from that calculated for a similarly doped plane diode because of a buildup in the electric field at the relatively sharp corners. For a shallow junction depth (1 μm) in lightly doped silicon (10^{14} atoms/cm³), such curvature can cause a 20:1 loss in breakdown voltage. High-voltage diodes use relatively deep junction depths (approximately 20 μm); as a result, a 3:1 loss is more typical. Special techniques (such as field plating over stepped oxides) can raise the breakdown voltage to approximately 70 % of that of the plane diode.

Curvature-limited breakdown mechanisms cause the relatively shal-

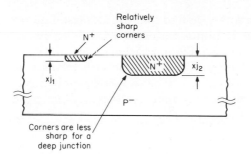

Fig. 2-9. Junction curvature in planar junctions.

low (0.2 μm) P-type ion-implanted resistors (to be described shortly) used in linear ICs to have a breakdown-voltage limit (P-type resistor to N-type epi) of approximately 40 V. The deeper junction depth (2.5 μm) of a P-type base-diffused resistor allows a reverse voltage of 100 V.

Zener Breakdown

When impurity doping on both sides of the junction is increased, the SCL is reduced in width. For reverse-biasing of a P^+N^+ diode, a different breakdown mechanism occurs: the true zener breakdown. This is the same mechanism that we previously found finally limits the voltages that can be applied to good crystalline insulators. The strange fact is that zener breakdown requires a higher value of electric field (approximately 10^8 V/m, or 100 V/μm) than avalanche breakdown. The obvious problem is how can we have a more intense field than that needed for avalanche breakdown and yet not be in avalanche breakdown? The answer is that the SCL in these heavily doped diodes is *so narrow* (200 to 500 Å) that the thermally generated carriers *cannot be sufficiently accelerated* in such short distances to smash into the lattice with enough energy to start the avalanching process. What happens instead is that the more intense electric-field strength allowed under these circumstances directly breaks up the covalent bonds of the lattice to obtain electron-hole pairs and thereby support a current. Current carriers no longer need to be accelerated by the field. *True zener breakdown*, therefore, only takes place in more heavily doped diodes (both sides must be heavily doped), and this limits these diodes to the 2- to 5-V range. (Physicists also refer to this as *tunneling*, which we will consider later in this chapter when we discuss the *tunnel diode.*)

Zener breakdown is essentially noise-free, and if we increase the ambient temperature, we find that the increased thermal energy actually makes the covalent bonds easier to break. This provides a negative *TC* of breakdown voltage. For breakdowns in the 5- to 6-V range, both breakdown mechanisms are simultaneously taking place, and this is why the standard 5.1-V *zener diode* has approximately a 0 *TC*, a *naturally* temperature-compensated reference voltage.

Ion Implantation and the Buried Zener.   For standard, diffused avalanche diodes, critical breakdown occurs where the PN junctions are the more heavily doped—at the surface. High electric fields at the surface cause undesirable increases in the noise voltage of a diode and tend to create changes in the magnitude of the breakdown voltage over time. To eliminate these problems, an ion-implanted subsurface avalanche diode has been used in linear IC voltage regulators and voltage references.

Ion implantation is a way to get dopants into silicon; implantation is accomplished by bombarding the surfaces of silicon wafers with electrically charged dopant atoms that are accelerated by an electric field. Ion implantation takes place within a vacuum chamber. The penetration of the ions depends on the energy of the ions and the orientation of the crystal structure of the silicon wafers. For example, if the implanting beam is parallel to a major crystal axis, or plane, the ions can penetrate very deeply. However, better reproducibility is obtained if the ion beam is misaligned with respect to the axes of the crystal.

The use of ion implantation can create a richer doping level below the surface, which places the breakdown region of the avalanche diode in the bulk, as shown in Figure 2-10. When diffusion furnaces are used

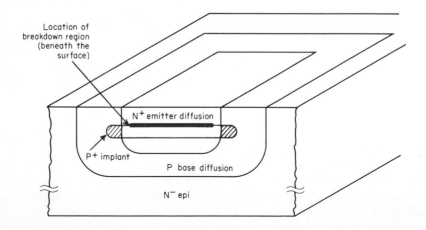

to create a doped region in silicon, the largest doping density occurs at (or very near) the surface. Ion implanters therefore allow the fabrication of junction-doping profiles that are impossible to obtain with diffusion techniques. Moving the more heavily doped regions below the surface and creating the breakdown within the bulk produces less noise and smaller voltage changes over time than are obtainable with standard surface-breakdown diodes.

Before we leave this section, let us take a short look at another type of breakdown.

Thickness-Limited Breakdown

The mechanism involved in thickness-limited breakdown is important to the designers of semiconductor products. This type of breakdown occurs in high-voltage devices when the SCL stops spreading with increases in the applied reverse-bias voltage or when the SCL touches a contact, which causes a breakdown-like current to flow. Breakdown does not occur if the SCL simply spreads into a more heavily doped region of the same doping type. However, when this type of spread occurs, the SCL does not increase as rapidly on the heavily doped side of the junction. Therefore, the electric field builds up, because the separation of the SCL is increasing less rapidly. Generally, increasing the reverse-bias voltage also increases the spread of the SCL, and the electric-field strength (the ratio of these two) therefore increases relatively slowly. Unexpected premature breakdown voltages can result when the SCL essentially stops spreading. Obviously, this is an important factor in the design of high-voltage devices.

2.4 UNDERSTANDING DIODES

We have covered many of the reverse-biased characteristics of diodes. Now we will consider forward-biasing and the mechanisms and special types of diodes based on it.

Forward-Biasing

We have seen how externally applied reverse voltages can increase the magnitude of the built-in electric field. Now we will consider what happens when this built-in electric field is reduced by the application of a forward-biasing voltage, as shown in Figure 2-11. If we remember that this built-in field is responsible for an upper limit on the total number of charge carriers allowed to initially diffuse across the junction, we

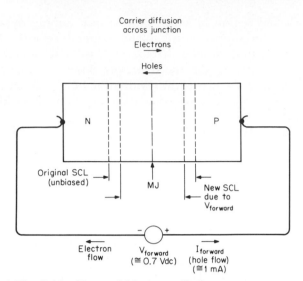

Fig. 2-11. Forward-biasing a diode.

can see that anything that reduces the strength of the field will allow a continuous, steady diffusion of charge carriers across the junction. This is what occurs in a PN junction under *forward-biasing,* or the *conducting condition.* Relatively small changes in the forward-biasing voltage can cause relatively large changes in the dc forward current of the diode. (Later we will see that this is the basic reason for the large transconductance of bipolar transistors).

The details of this forward current are shown in Figure 2-12. The SCL is exaggerated in the figure to make it visible. In reality, SCL width is 2 to 3 orders of magnitude thinner than the regions containing the injected minority carriers (which are shown just outside the SCL on both sides of the junction).

As the carriers diffuse into or are "injected" into the region on the opposite side of the junction, they become minority carriers. As a result, their concentration is exponentially reduced as they move through the region. This reduction in minority carriers is due to recombination with majority carriers, which removes both carriers, (mutually annihilates them).

Because the diode in the figure is an N^+P^- diode, there are more electrons than holes crossing the junction. Most of the forward current is therefore due to electron flow. In general, forward current of a diode is the sum of both electron and hole current flow.

The maximum density of electrons occurs at the edge of the SCL

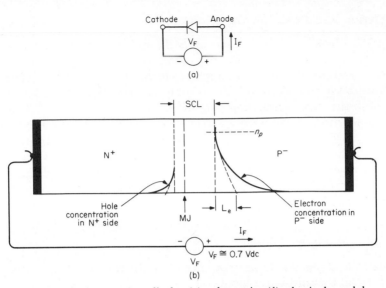

Fig. 2-12. A conducting diode: (a) schematic; (b) physical model.

on the P-type side of the diode. This is shown as n_p in the figure, and n_p increases exponentially with the magnitude of the external forward-biasing voltage. Larger values of ON voltage will therefore raise this maximum electron density because more electrons are allowed to diffuse across the junction. Remember, in semiconductor operation, *diode forward-biasing voltages are associated with* injected minority carrier concentration, (n_p).

For the diode in the figure, the magnitude of the dc forward current flow is established by the initial slope of the electron concentration. The linear extrapolation of this initial slope also provides the value of the *diffusion length* L_e of the electrons in the P-type material. This is a measure of how far the minority carriers can penetrate without being recombined with majority carriers. The forward current of this diode is the diffusion current that results from the minority-carrier concentration gradient. This gradient is proportional to both the maximum minority-carrier concentration and the diffusion length. Therefore, forward current also depends on both, or

$$I_F \propto \frac{n_p}{L_e}$$

A diode with more equal doping levels on both sides will have an extra term (p_n/L_h) in the expression for forward current; this is due to the other carrier type.

Depletion and Diffusion Capacitance

A forward-biased diode still has an SCL, and therefore, a depletion or junction capacitance exists in association with the immobile charge embedded within the lattice. To rapidly change the conduction state of a forward-biased diode, one must supply enough charge to the SCL to establish the new conduction conditions.

In addition, the minority carriers existing just outside the SCL on each side of the junction also must be changed. This occurs relatively slowly because we have to wait for carriers to diffuse or recombine. In models of diodes and transistors, another capacitor, *the diffusion capacitance,* is used to account for this phenomenon. However, it applies only to a forward-biased junction.

High-Conductance Diodes

Some special silicon diodes conduct large forward currents for a relatively small forward-biasing voltage. The basic mechanism of these *high-conductance,* or *thin-base, diodes* is to locate the metal contact on the P-type side closer to the junction than the diffusion length. This will increase the slope of the electron concentration gradient by, in effect, reducing the "apparent" value of L_e, as shown in Figure 2-13. We can

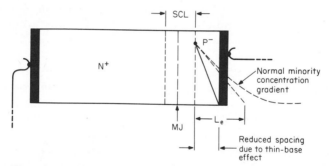

Fig. 2-13.   The basis of the high-conductance diode.

see in this figure that the metal contact on the P-type side has been moved closer to the junction than one diffusion length. Therefore, for the same forward-biasing voltage (the same n_p), we will get a larger diffusion current flow and therefore obtain a higher-conductance diode. As expected, these diodes have a built-in thickness-limiting breakdown mechanism that restricts the reverse voltage that can be applied to them.

High-conductance diodes are also useful for fast-switching applications because the total minority-carrier (electron) concentration is reduced for a given operating current. Therefore, when this diode is switched from ON to OFF, less time is needed for the electrons to recombine.

The operation of bipolar transistors naturally follows this thin-base (or narrow-base) diode concept. Before we discuss bipolar transistors, let us take a look at some unusual diodes.

Backward Diodes

Useful and interesting diodes result from raising the doping levels on both sides of the junction. For heavy N^+ doping and moderately heavy P-type doping, a diode results that is in breakdown for a small value of externally applied reverse voltage. These diodes make useful detectors or rectifiers for low-voltage signals because the *reverse* characteristics appear like the *forward* characteristics of a diode with an unusually small forward-voltage drop (Figure 2-14). This *backwardness* of operation pro-

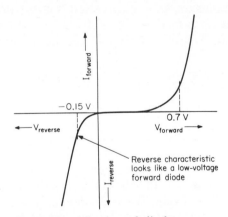

Fig. 2-14. A backward diode.

vides the name for this diode. *No minority-carrier storage exists* because this apparently forward-conducting diode *is really in breakdown*. Consequently, these diodes are capable of operation at high frequencies.

Quantum-Mechanical Tunneling and the Tunnel Diode

When a diode has heavy N^+ doping and a more heavily doped P-type region, it is called an *Esaki*, or *tunnel, diode*. (Dr. Leo Esaki shared the Nobel Prize in physics in 1973 with Ivar Giaever for research in tunneling in relation to semiconductor and superconductor electronics.) As mentioned earlier, the zener mechanism also involves tunneling, but zener diodes require an external reverse biasing voltage to cause tunneling to occur. With heavier doping on both sides of the junction, a diode results which is naturally in breakdown. The built-in field in the resulting very narrow SCL is large enough to cause inherent breakdown. Therefore, if heavier doping is used in an attempt to fabricate a low-voltage (less than 2-V) zener diode, a new diode results: the tunnel diode.

This tunneling mechanism is indeed unusual. It is one of the phenomena that required a departure from the "charged billiard ball" model of the electron and the classical treatment of the particle nature of the electron.

Quantum physics was needed. First elucidated by Max Planck at the turn of the century, quantum physics was used by Albert Einstein 5 years later to explain the photoelectric effect. In 1913, the Danish physicist Niels Bohr extended Planck's ideas to the atom, which initiated a relatively rapid series of discoveries in atomic physics. In 1925, the French physicist Louis de Broglie extended Bohr's work and introduced a "pilot-wave" theory, which in the next year was bolstered by the Austrian physicist Erwin Schrodinger to become the theory of wave mechanics.

The famous Schrodinger equations had probabilistic solutions that did not exist in Newtonian physics. One such discovery was the quantum-mechanical tunneling of a barrier by an electron. Essentially the explanation is this: if the barrier is thin enough, a certain probability exists for the transmission of the electron through the barrier, even though the energy of the electron is less than the potential of the barrier.

To better understand the quantum-mechanical tunneling of a barrier by an electron, we can consider a similarly strange phenomenon that takes place with a light beam. If we imagine a light beam *shining within a glass plate,* we can approach an understanding of quantum-mechanical tunneling by considering how the light beam is reflected at the glass-to-air interface. If the light beam is perpendicular to the interface, light will escape into the air. However, if the light beam is more nearly parallel to the interface, the light will be reflected at the interface and will remain within the glass.

From the particle nature of light (similar to the particle nature of the electron), we are led to think that reflection of the light-beam particles (the photons) at the interface is the same as the bouncing (or reflecting) of a table-tennis ball off the surface of a table. According to the wave theory of light, what is *really* happening is that these totally reflecting light waves *actually penetrate* into the air to a depth (or height) of several wavelengths and then are *thrown back* into the glass. This has been proven by bringing a second glass plate near the first, within a few wavelengths (for visible light, this is a few microns). Some of the reflected light can be captured by the second glass plate and will continue to propagate within the second plate, in the original direction of the light beam in the first plate.

In a similar manner, electrons (considered to move in a wave pattern) can *tunnel through* a barrier, but only if the width of the barrier is extremely thin. In fact, tunneling occurs only in the narrow SCLs that exist within N^+P^+ diodes. This is the basis for the previously described zener diodes as well as the tunnel diode, where the SCL is less than 50 Å wide. Therefore, the "soft" breakdowns associated with the low-voltage zener diodes result from tunneling currents.

Tunnel currents can adjust much faster to changes in the applied voltages because they do not depend on the relatively slow diffusion and recombination mechanisms normally associated with diode current flow. Consequently, the tunnel diode is capable of operation at very high frequencies.

Tunnel diodes are in breakdown at 0V. The rather strange thing is that *they remain in breakdown for* some small amount of *forward-biasing voltage.* As the forward voltage is increased, the diode slowly *comes out of breakdown,* causing the *current to decrease as the forward voltage increases.* This provides the *negative-resistance region* of these diodes (Figure 2-15), which has allowed them to be used as amplifiers and oscillators. The dotted curve in Figure 2-15 shows the normal forward characteristics of a diode. The tunnel diode *appears normal* at higher values of forward-biasing voltages.

The power available from the negative-resistance characteristic of a germanium tunnel diode is limited because the extent of the negative-

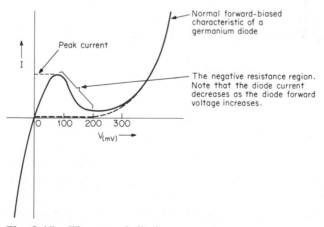

Fig. 2-15. The tunnel diode.

resistance region is less than a few hundred millivolts. Manufacturers have therefore worked to raise the peak currents to increase the available power.

Tunnel diodes can be operated at microwave frequencies and, as a result, are rather difficult to work with because they can (and will) *oscillate* at many different frequencies *simultaneously*. Special circuits use one tunnel diode for the radio-frequency amplifier, local oscillator, mixer, and first intermediate-frequency amplifier *all at the same time*.

P-i-N and Step-Recovery Diodes

As might be expected, P-i-N diodes consist of two oppositely doped silicon regions separated by a layer of intrinsic, or undoped, silicon. When this diode is reverse-biased, the two regions of fixed charge are separated by the relatively *long intrinsic silicon region,* which acts as an insulator. This provides *the most effective structure for supporting a high reverse voltage* and greatly reduces the capacitance of the diode. When this device is turned ON (Figure 2-16), the intrinsic center region is flooded with

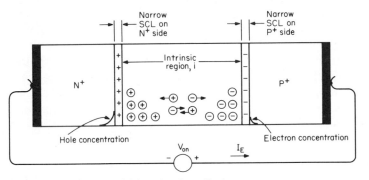

Fig. 2-16. A forward-biased P-i-N diode.

charge carriers of both types. The presence of these mobile carriers *conductivity modulates* (reduces the resistance of) the intrinsic layer, which allows the diode to function as a switch in microwave applications.

Notice that the heavily doped P^+ and N^+ regions reduce the minority-carrier concentrations outside the SCL. Most of the minority charge therefore exists in the intrinsic region. A new phenomemon now takes place: the uncovered charge in each SCL at the heavily doped ends tends to limit the number of mobile minority carriers of this polarity injected into the end regions. The rate of injection of the carriers is established by the dc current that is flowing. In Figure 2-16, the mobile holes near the N^+ region and the mobile electrons near the P^+ region have no arrows on their symbols, which indicates that they are not moving into the end regions. This is different from a regular diode; as the applied voltage is rapidly reversed, the new electric-field direction rapidly removes the charge carriers from the i-region. This mechanism is used in the *step-recovery*, or *snap, diode.*

In the snap application, the P-i-N diode has a narrower i-region. The snap diode is first *charged* by operating it under a forward bias. This floods the i-region with mobile charge carriers. Then the current is reversed, which removes the charge in the i-region. During the relatively short time it takes to remove the charge, the reverse voltage across the diode remains relatively low in value. When the diode recovers, its low capacitance allows the reverse voltage to *snap* up to the full value of the applied reverse-biasing voltage *in an extremely short time* [approximately 10 picoseconds (ps) per micrometer (μm) of i-region length, or 5 to 50 ps for a typical diode]. Consequently, snap diodes can increase the rise time of a pulse or even "square up" a sine wave. This latter use provides harmonics, which are useful for frequency multipliers.

Schottky Diodes

A metal layer deposited on a lightly doped N-type silicon material creates a *Schottky-barrier*, or *hot-carrier, diode.* When the metal and the semiconductor are properly selected, the semiconductor becomes the N-type region of the resulting diode. This can be thought of as *a half-diode because only one side is a silicon region.* As in a normal PN diode, electrons diffuse from the N-type semiconductor, but in a Schottky diode they form a thin, dense layer at the interface of the metal and the semiconductor (Figure 2-17). The semiconductor will have an SCL that extends into

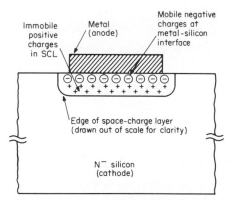

Fig. 2-17.   Enlarged view of a Schottky diode.

it and uncovers fixed positive charge in the lattice. The resulting electric field limits electron diffusion from the semiconductor to the metal.

A Schottky diode follows an exponential *V-I* curve. The forward current is carried by the electrons that flow from the semiconductor to the metal contact. Recombination is no longer involved because the injected electrons are majority carriers in the metal. Therefore, these diodes can switch from ON to OFF in a few picoseconds, which has made them useful for high-frequency detectors. The forward-voltage drop can be kept low (0.3 V), which makes them efficient power-supply rectifiers. In addition, Schottky diodes are useful for shunting parasitic silicon diodes in ICs. By conducting at lower forward voltages, they prevent conduction of the protected silicon diode within the IC.

Schottky diodes are also used in this current-shunting manner to prevent saturation in modified T²L logic circuits (we will look at this in Chapter 5). In addition, they have been used to modulate the channel region in a new form of high-frequency metal-semiconductor junction field-effect transistor (these MESFETs are considered in Chapter 7).

Further, the Schottky diode acts as the basis of the *ohmic connection,* which is needed when a surface-metal layer must be contacted to a doped region of silicon. For large doping levels in the silicon region (either P⁺ or N⁺), the resulting SCL penetrates very little into the doped semiconductor. Such a thin SCL *is easily tunneled* by electrons, even with essentially no external field. This is the mechanism of the *ohmic contact,* which can be thought of as *a tunneling Schottky diode.* These nonrectifying contacts are used to connect to the transistors, diodes, and resistors in ICs.

The Ideal-Diode Equation

Since bipolar transistor operation is based on the PN diode, we will look more closely at the quantitative relationship between voltage and current in a diode. This is given by the ideal-diode equation:

$$I = I_S (\exp \frac{V_D}{V_T} - 1) \tag{2.1}$$

where $\exp \dfrac{V_D}{V_T} = e^{\,V_D / V_T}$

I_S = the reverse saturation current

V_D = the external biasing voltage of the diode (assumed positive for forward bias)

V_T = the thermal voltage (26 mV at room temperature)

Thermal voltage can be calculated from the following equation:

$$V_T = \frac{kT}{q}$$

where k = Boltzmann's constant (1.38×10^{-23} joules per °K)

q = the charge of an electron (1.6×10^{-19} coulombs)

T = absolute temperature in degrees Kelvin [T (°K) = T (°C) + 273]

Some errors can result from placing too much faith in this equation. In general, such errors are acceptable; it is only in more exact modeling of forward currents or leakage currents that trouble arises. For example, the ideal-diode equation predicts that the reverse leakage current would rapidly reach a constant or reverse saturated value (equal to I_S and therefore the name for this term) for values of V_D that are negative and large compared with V_T (26 mV). Note that for reverse bias this equation becomes

$$I = I_S \left[\frac{1}{(\exp |V_D|/V_T)} - 1 \right]$$

which equals $-I_S$ because of the negative sign of V_D, which causes the value of first term in the brackets to rapidly decrease in comparison with that of the second term (-1). In fact, actual diode leakage current *increases slightly* with further increases in reverse-biasing voltage (and real diodes eventually break down as the reverse voltage is increased). The voltage-dependent current increase results from both surface leakages and the increased width of the SCL. The parameter I_S only accounts for leakage currents generated within a few diffusion lengths outside the edges of the SCL. It does not account for carrier-generation effects within the SCL. More thermally generated electron-hole pairs within this wider SCL will increase the leakage current.

A final caution concerning the use of the ideal-diode equation is that the *V-I* characteristic of real diodes may depart from this simple model at both low currents and high currents, where the actual relationship can still be predicted by the ideal-diode equation if a multiplier m is added to V_T, as mV_T, where $1 \leq m \leq 2$. Bulk resistance in the diode fabrication also causes a shift away from the fast-changing exponential equation to a linear relationship at high currents.

Because the theoretical leakage current is equal to I_S, any changes in this parameter directly affect the magnitude of the leakage current. Since I_S varies directly with the junction area of the diode, larger diodes have larger leakage currents.

The forward characteristics of a diode are more dominated by the fast-changing exponential term and are less dependent on the exact value of I_S. For example, doubling the junction area of a given diode will double I_S, but will only reduce the forward ON voltage by 18 mV out of, say, 600 mV. Therefore, a 100 % increase in I_S gives a 3 % reduction in V_D. We will look at the V_D (or V_{BE}) versus I_D (or I_C) relationship more closely in Chapter 5.

A plot of this ideal-diode equation is given in Figure 2-18, and much

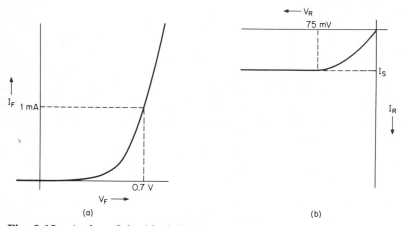

(a)   (b)

Fig. 2-18. A plot of the ideal-diode equation: (*a*) forward characteristic; (*b*) expanded reverse characteristic.

about semiconductors can be derived from this relatively simple relation. Because the magnitude of forward currents is much larger than that of reverse currents, this equation also has been plotted in a separate graph (Figure 2-18*b*), so that a change in the scaling of the vertical current axis can be made. Note that the diode forward current increases exponentially with applied voltage. This causes relatively large changes in diode current for small changes in forward voltage.

The reverse saturation current I_S undergoes large changes with temperature. It approximately doubles for every 10°C increase in the temperature of the diode. This makes the *V-I* characteristics of a diode exceedingly temperature-dependent in both the forward and reverse modes. For a fixed forward-biasing voltage, the diode current would therefore rapidly increase as temperature increases, or alternatively, the amount of forward voltage necessary to provide a given forward current would fall as temperature increases. This is the basis of the −1 to −3 mV/°C temperature coefficient of the forward voltage of a diode.

Since actual measured leakage currents are several orders of magnitude (approximately 100 times) larger than I_S, a value for I_S in the ideal-diode equation *for forward-current calculations* should be determined from a *forward-current and voltage measurement* and not from a leakage-current measurement. One value for I_S *will not work for both* forward and reverse conditions.

ON Resistance of a Diode

When a diode is used in circuit designs, the question of what the ON, or conducting, resistance of the diode is must be answered. For large signals, such as in a power-supply rectifier application, the diode makes large resistance changes from ON to OFF.

The resistance a diode presents to a low-level, or small-amplitude, signal can be determined from the ideal-diode equation. Note that the slope of the diode characteristic of Figure 2-18 is constantly increasing and is therefore not a single value, but rather is current- or voltage-dependent. The slope of this curve can be determined graphically by drawing a tangent to the curve at the operating point in question. The dimensions of this slope are units of current (in milliamperes, mA) per volt, or conductance (the reciprocal of resistance).

The slope can be calculated by taking the derivative of the diode current with respect to the diode voltage for the forward-biased diode (for those without a calculus background, this insight will not be obvious, but the simple equation that results is still useful). This derivative, or small-signal diode conductance g, becomes

$$g = \frac{dI_D}{dV_D} = [I_S \ (\exp \ V_D/V_T)]\frac{1}{V_T} \tag{2.2}$$

Where the factor in brackets is the value of the diode current for a forward-biased diode [since $(\exp \ V_D/V_T) \gg 1$ for a typical forward bias, the -1 can be dropped in comparison to simplify the diode equation 2.1]. Substituting this into Equation 2.2 provides a simpler form:

$$g = \frac{I}{V_T}$$

The small-signal resistance r of the diode is the reciprocal of g and is therefore given by

$$r = \frac{1}{g} = \frac{V_T}{I}$$

At room temperature (25°C, or 298°K), this becomes

$$r = \frac{26 \times 10^{-3}}{I}$$

or

$$r = \frac{26}{I \ (\text{mA})} \ \text{ohms}$$

if I is expressed in milliamperes (mA). This shows that a diode conducting a forward current of 1 mA has a small-signal resistance of 26 ohms (Ω).

Note the importance of the concept of *small-signal resistance*. This is a way of stating that a *single value of resistance* for a diode *only applies* if the *alternating-current (ac) voltage* swing across the diode is of *small magnitude* (approximately 5 mV peak). For larger ac voltages, the incremental resistance of the diode is changing as the voltage changes; this is why diodes are highly nonlinear circuit elements.

Commercially available small-signal diodes (such as the 1N914) can have approximately 3 Ω of bulk resistance associated with their construction. Even under large forward-current operation, these diodes will not have a resistance lower than this 3-Ω limit. For small values of biasing current, say 1 μA, the small-signal resistance is

$$r = \frac{26}{0.001 \ (\text{mA})} = 26 \ \text{k}\Omega \ (\text{kilo-ohms})$$

Note that a wide range of resistance values can be obtained with a diode, and unless the ac signal magnitudes are kept small, large nonlinear changes in the diode resistance will result.

2.5 OPTOELECTRONIC DIODES

Optoelectronic diodes can be of two basic types. Those that create light are called *light emitters,* and those that respond to light are called *light detectors.* There is much present-day research in this area in an attempt to replace the television camera and the viewing tube of a TV set with solid-state optoelectronic devices. Here we will consider only the more common opto devices, the *photodiode* and the *light-emitting diode.*

Photodiode Light Detectors

The photodiode responds to light energy (photons) that impacts on the silicon lattice near and within the PN junction. The frequency of the light must be in the correct range, which means that the energy of its photons must be sufficient to break covalent bonds. For silicon, light in the near-infrared and visible regions of the spectrum will cause such a photovoltaic response (broken bonds, which provide the carriers for current flow).

The silicon crystal presents a barrier to the passage of light. In fact, penetration of the light into the silicon decreases exponentially with distance into the crystal. Further, more energetic, higher-frequency photons suffer more attenuation than do those of lower frequencies. Therefore, photodiodes have a particular spectral response that depends on the diode junction depth. For example, blue light is attenuated such that only 37 % of its intensity at the surface remains at a depth of 1 μm and only 5 % remains at 3 μm. As the PN junction is moved closer to the surface of the silicon, the response to the blue end of the spectrum is increased. Deep junctions will have little or no response to high frequencies and will therefore provide a red shift in the response of the photodiode.

When the photon energy is less than the band-gap energy (1.1 eV for silicon), the photons are not absorbed and can pass through the crystal; silicon therefore becomes *transparent* to *infrared*. When the photon energy is equal to the band-gap energy, the photons are absorbed. Photons with greater energy (wavelengths less than 1100 Å) are still absorbed, and the excess energy is dissipated within the silicon as heat.

You will recall that as a result of thermal energy, a reverse-biased diode will conduct a leakage current. This is due to electron-hole pair generation in and near the junction. The photoresponse of a diode is similar to leakage current (which, in this case, is called *photocurrent*) that results when electron-hole pairs are generated by the energy of the photons striking the lattice. Thus there are two energy sources generating "leakage" currents, and the minimum dark response of the photodiode is limited by the ever-present thermal component, now called *dark current.* This can be minimized by operating the diode at 0 V (for example, the *virtual ground* of an op amp, as shown in Figure 2-19)

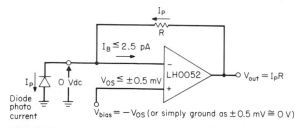

Fig. 2-19. Measuring photo current with a diode voltage of 0 V. V_{OS} is the offset voltage (ideally equal to 0 V) of the particular op amp chosen.

and keeping it at as low a temperature as possible. This fixed-voltage biasing also will speed up the response of the diode to rapidly changing light conditions, because the feedback action of the op amp will require only a small (and therefore less time-consuming) voltage change across the (undesired) junction capacitance of the diode. For the highest-frequency applications, large values of reverse-biasing voltages are used to reduce the junction capacitance of the diode as well as to increase the volume of the SCL. This also increases the percentage of the total photocurrent, owing to the higher-frequency SCL component (where the high electric-field strength moves carriers very fast) and reduces the contribution from the slower diffusion components of the photocurrent generated just outside the SCL.

Photodiodes are used as light receivers in optical transmission systems. When compared with wire communications, the use of optics greatly increases the number of signals that can be simultaneously present on one communication channel (or increases the signaling rate for a single signal). Consequently, light pipes (small glass or plastic tubes that provide the "wires" for an optical path) are being investigated for both their use in the telephone industry and the computer industry. Further, optical transmission eliminates the requirement for a common electric ground connection and also eliminates the problems of electric interference with the signaling path. These are important features in many electronic systems.

Light-Emitting Diodes

The more efficient and rugged solid-state light-emitting diodes (LEDs) have essentially replaced the older incandescent varieties of panel lamps. This direct method of light generation allows LEDs to be switched ON and OFF at relatively high speeds (approximately 10 nanoseconds). These diodes make use of the carrier recombination that takes place in a forward-biased diode.

Recombination can be either direct or indirect. With direct recombination, a conducting electron drops back to the valence band and loses all its excess energy. The energy difference, or band gap, is large enough in silicon (1.1 eV) to cause a photon of light to be emitted from the crystal for each direct recombination. However, the probability of a direct recombination in silicon is relatively small, and the energy of the conducting electrons is usually lost in two stages or steps through use of recombination centers (places, or allowed energy levels, for electrons) within the band gap. Each of the step losses in energy is not enough to emit a photon of light. Therefore, the energy heats the crystal because it is transferred to the lattice. This is called *phonon generation* (acoustical coupling to the lattice).

Special semiconductor compounds that encourage direct recombination can be made, and these are used in LEDs. The common electroluminescent semiconductor material for infrared applications is gallium arsenide (GaAs). GaAs has been mixed with other materials, such as gallium phosphide (GaP), to shift the energy of the recombining electrons and thereby provide orange and most of the standard red LEDs. Gallium phosphide also has been used separately for both red and green LEDs.

The light intensity of all these diodes increases with increases in the forward current. This is due to the increased recombination that results from larger forward-current flow. Typical LEDs have a relatively large forward voltage (approximately 1.5 V_{DC}) and are operated at a current of approximately 10 mA. If a special physical construction is used, the light radiation can be made coherent (a single-frequency, or monochromatic, light), and this is the basis of the injection laser.

The compound semiconductors used to make LEDs, such as gallium arsenide (sometimes called a *semi-insulator* because of its larger band-gap energy, 1.4 eV), are doped in a slightly different way than silicon. Zinc, with two valence electrons, is used to substitute for gallium (which has three valence electrons) to provide a P-type dopant. Similarly, sulfur, with six valence electrons, is substituted for arsenic (which has five) for an N-type dopant.

These compound semiconductors are harder to process because *they will decompose* at elevated temperatures. For example, if GaAs is placed in a diffusion furnace, the As tends to *boil off*, and this changes the composition of the material. This was not a problem with single-element semiconductor materials, such as germanium and silicon. Lower-temperature *epitaxial techniques* (which we will consider in Chapter 4) are therefore used for the *generation of the doped regions* in GaAs LED products. Consequently, bipolar transistors are difficult to fabricate with this material, but a great deal of research is underway to obtain the higher frequency and higher operating temperature capabilities of GaAs transistors when compared with silicon transistors.

CHAPTER III

An Intuitive Description of Bipolar Transistor Operation

With the background provided by the discussions of the PN diode, we can now readily understand the operation of bipolar transistors. Most of the peculiarities and characteristics of bipolar transistors can be modeled and understood by simple considerations of the basic physical mechanisms of semiconductors. The word *bipolar* indicates that both majority and minority carriers are involved in the functioning of these transistors.

3.1 TRANSISTOR BASICS

The name *transistor* was coined by J. R. Pierce of Bell Laboratories from *transfer resistor*. This suggests that transistors can be thought of as making use of an input current to control an output voltage. Confusion often results with the identification of the proper input parameter for a transistor, since voltage, current, and even charge have been used. This is in contrast to vacuum tubes, in which the input parameter is almost always voltage.

We can jump right to the essence of transistor action by noting the strong similarity between transistors and thin-base diodes. The only difference is that instead of locating a metal contact close to the emitting junction, a second reverse-biased diode is brought close to the junction. When this is done, the SCL of the second diode sweeps up or *collects* most of the minority carriers provided, or *emitted*, by the first junction. This is why the names *emitter* and *collector* are used for two of the leads of a transistor. We will consider the origin of the name *base* for the third lead in the next chapter.

The main idea of the transistor is that the forward current of the *first*, or *emitter-base, junction* is stolen (or collected) by the *second*, or *collector-base, junction*. An input (or control signal) is used to get the emitter-base diode to conduct, and the resulting current from this forward bias "mysteriously" *appears* in the output, or collector-base, circuit.

It might be puzzling to the student why one cannot make an NPN transistor simply by connecting two 1N914 discrete diodes in series such that the common anode point serves as the base and the cathodes become the emitter and collector, respectively, as shown in Figure 3-1. Hopefully, the previous discussions of the PN diode will bring to

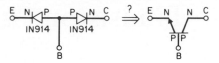

Fig. 3-1. Why not make a transistor out of two diodes?

mind the fact that the holes injected into the N-type side and the electrons injected into the P-type side *have all recombined* before they reach the metallized contacts of the diodes. Thus, with two discrete diodes, *there are no minority carriers* available for the second diode to steal. (Transistor action could result if, for example, both P-type regions were thinned down and then placed in contact with each other and a base lead attached to this joint, although this would not be a very good transistor.)

Transconductance of a Transistor

Transconductance, like conductance, is still the ratio of a current to a voltage. However, in transconductance, the ratio is between an *output* current and an *input* voltage. Consequently, transconductance in a transistor is an indication of how a small change in input voltage is transferred to the output to cause a change in the output current of the device, or

$$g_m = \frac{\Delta I_C}{\Delta V_{BE}}$$

Therefore, g_m is a quality factor in an amplifying device, and large values of g_m are usually desired.

Because the collector steals essentially all the emitter current, the transconductance of a transistor is approximately the same as the conductance of a diode (actually, because of the thin-base effect, transistors have a larger value of transconductance), or

$$g_m = g_d = \frac{I_E}{V_T}$$

This occurs because a forward voltage applied to the base-emitter diode of a transistor will cause a normal-diode forward current to flow.

The structure and basic operation of a transistor causes most of this resulting forward current to be taken up by the collector. As a result, the current flow in the base lead is very small. This means that if a voltage is applied to the base, or input, lead of a common-emitter transistor, the diode forward current that results will appear in the collector (the output lead) of the transistor, as shown in Figure 3-2. For example,

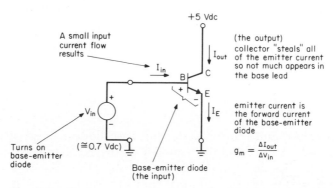

Fig. 3-2. Basic transconductance of a bipolar transistor.

if a transistor were operated at a collector current of 1 mA, g_m would be

$$g_m = \frac{1 \times 10^{-3}}{26 \times 10^{-3}} = 38 \text{ mA/V}$$

The units for g_m are actually *mhos* (ohms spelled backwards), but are even more meaningful if stated as the output current change per millivolt (mV) of input voltage change, because g_m is valid only for low-level or small-signal conditions. Therefore, millivolt input signals are expected, not volts. Multiplying both the numerator and the denominator of the previous expression for g_m by 10^{-3} changes the units of g_m to 38 microamperes (μA) per millivolt. This states that a 1-mV change in the voltage at the base-emitter junction, the input, would cause a change of 38 μA in the collector current, or output current, of a single-stage common-emitter transistor amplifier. When this changing collector current is flowing in a 10-kΩ collector load, or biasing resistor, a 380-mV change in output voltage would result. The voltage gain of this stage is $V_{\text{out}}/V_{\text{in}}$ or 380 (380 mV/ 1 mV). Large values for both g_m and the load resistor provide large values of gain. The maximum voltage gain which can be obtained with a single transistor amplifier is approximately 3000. We will consider this again later in this chapter after the concept of the output impedance of a transistor has been introduced.

Notice that the small value of base current, which results because the collector is stealing essentially all the emitter current, provides a higher input resistance at the base terminal than the simple ON resistance of a conducting diode. This is the basis for the differences in input resistance between the common-emitter and common-base connections.

In Chapter 2 we saw how thin-base diodes can have high conductance. It also happens that a wide-base-width transistor will have a reduced g_m. In general, transistors with wide base widths are not desired; usually they are parasitic structures in ICs, where they are useful, but not optimum, added devices. This is true of lateral PNPs in linear ICs and vertical NPNs in CMOS products (we will have more to say about these in Chapters 5 and 7, respectively).

Internal Transistor Action

Let us look more closely at the thin-base diode concept of a transistor (Figure 3-3). Figure 3-3*b* is a model of the NPN transistor shown in Figure 3-3*a.* In the symbol for a transistor, the emitter arrow shows the direction of the conventional (not electron) current flow. (A PNP transistor symbol should therefore be drawn with the emitter at the

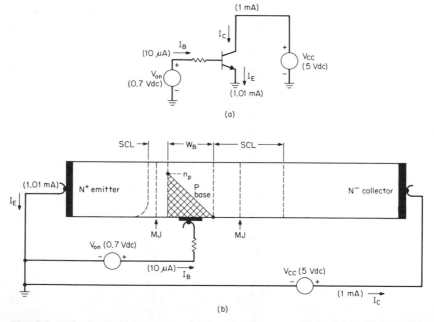

Fig. 3-3. The basics of transistor action: (*a*) the actual transistor; (*b*) a model of the transistor.

top of the schematic so the current flows from the top to the bottom of the page. This drawing convention makes circuits much easier to understand.)

To make things more definite, the figure gives representative values for terminal voltages and currents that typify normal transistor operation. In addition, a protection resistor has been added in series with the voltage source to accommodate the people who do not like to see voltage sources directly applied to diodes (theoretically it is no problem, but practically it is easy to burn out diodes during experimentation if a good voltage source is used without some form of current limiting). The emitter current (1.01 mA) is the sum of the collector current (1 mA) and the base current (10 μA, or 0.01 mA). The relative magnitudes of these currents show that most of the current emitted by the *emitter* is collected by the *collector*. The ratio of these currents is represented by α and is defined as

$$\alpha = \frac{I_C}{I_E}$$

For this example,

$$\alpha = \frac{1 \text{ mA}}{1.01 \text{ mA}} = 0.99$$

Note that this ratio can never reach unity ($\alpha \neq 1$); however, in a *good* transistor it can get close.

Another current ratio useful for circuit design is, as might be expected, represented by β and is defined as

$$\beta = \frac{I_C}{I_B}$$

For this example,

$$\beta = \frac{1.0 \text{ mA}}{0.01 \text{ mA}} = 100$$

Since β relates the output current I_C to the input current I_B, it is called *current gain*. A β value of 100 is typical for current gain.

The input impedance at the base lead of the transistor in Figure 3-3 is simply β times the impedance r_e of the forward-biased emitter-base diode, because this is the factor by which the base current is reduced (remember, reducing the base current means that the impedance associated with the base lead is increased). The input impedance of the common-base connection is simply the dynamic emitter resistance r_e. For this example,

$$r_e = \frac{26}{I_E \text{ (mA)}}$$

or

$$r_e = \frac{26}{1.01} \cong 26 \ \Omega$$

Therefore, the input impedance at the base lead is

$$Z_{in} = \beta r_e = 100(26) = 2.6 \text{ k}\Omega$$

Before we continue discussing the mechanisms of transistor action, let us return to the model of Figure 3-3b. Note the relative impurity-doping levels: the emitter is the most heavily doped, followed by the base region, and the collector has the lightest doping level of the three. These relative impurity levels cause the asymmetrical spreads in the SCLs depicted in the figure. The greater spread of SCL in the collector-base junction will cause the collector-base biasing voltage to affect the base width W_B. This *base-width modulation* problem will be considered shortly. In addition, the figure also shows a small amount of back injection by the base into the emitter region (which undesirably increases the base current).

The transistor is shown for forward-mode conducting or ON operation. This requires that the emitter-base junction be forward-biased and that the collector-base junction be reverse-biased. Representative values for these biasing voltages are 0.7 and 5 V_{DC}, respectively, as is shown in the figure. (Actually, the collector-base bias is 4.3 V_{DC} in this example. The 0.7-V_{DC} base voltage must be subtracted from the collector-emitter voltage to provide the collector-base reverse-biasing voltage.)

The *effective base width* W_B is the difference between the edges of both SCLs and is shorter than the diffusion length of the minority carriers (electrons) in the P-type base region. In this respect, the transistor is similar to the thin-base diode previously described.

Factors Providing Undesirable Base Currents

"Good" transistors have small magnitudes of base current. To comprehend what makes an ideal transistor, we will consider the three major reasons for the existence of base current (large-signal effects will be considered in Section 3.3):

1. *Poor emitter efficiency.* This is the ratio of the forward currents which cross the emitter-base junction. Note that this diode is N^+P^- to favor

electron diffusion. The fact that some holes are *back-injected* into the emitter provides a component of the total base current (shown in Figure 3-3*b*).

It is interesting that a richly doped N^+ region is not completely ionized at room temperature (every dopant atom is *not* supplying an electron). When a silicon transistor is operated at higher junction temperatures, more of the emitter dopants supply electrons. This improves the emitter efficiency and is the principle reason that β increases by approximately 1 % per degree centigrade.

2. *Recombination in the base.* The presence of injected electrons, which are now minority carriers in the base region, gives rise to recombination with the holes that naturally exist in this P-type base region. All the injected emitter current is therefore not "transported" to the collector: the fraction that is successful is given by the *base-tranport factor.* This loss from recombination creates a second component of the total base current. This can be made smaller by using lighter doping levels for the base region of the transistor and reducing the base width.

Such recombination also causes the actual minority-carrier concentration to be slightly bowed downward from the straight line shown in Figure 3-3*b* and increases the mathematical complexities of exact analysis. For our purposes, we will neglect this recombination. As a result, the slope of the minority-carrier concentration in the base region was simply shown as a straight line that runs from the edge of the emitter-base SCL (where the electron concentration is a maximum of n_p) to the edge of the collector-base SCL (where the electron concentration is zero).

3. *Base current owing to recombination in the emitter-base SCL.* In silicon there are traps, or recombination centers, within the emitter-base SCL that increase the recombination of the injected electrons during their passage through this region. In addition, increased recombination can exist at the silicon surfaces (this can be greatly increased if the base-emitter diode has ever been operated in reverse-voltage breakdown). These factors cause excessive base currents at low levels of collector current, which reduces β. As the collector current is increased, the larger magnitude of base current will tend to dominate these smaller recombination currents and β will therefore increase.

As was mentioned in the discussion of diodes, the maximum electron concentration n_p depends on the base-emitter ON voltage biasing level. As shown in Figure 3-2*b*, the point at which the electron concentration goes to zero (the edge of the collector-base SCL) is determined by the reverse-biasing voltage of the collector-base diode. Consequently, the two end points of the straight line that represents the electron (or minority) concentration gradient can be related to the external biasing voltages applied to the transistor. The slope of this line determines the diffusion flow of the minority carriers across the base width. This will become the collector current, or

$$I_C \propto \frac{n_p}{W_B}$$

One final point is that the base current of a transistor is usually considered to result mainly from recombination within the base; therefore, base current depends on the total amount or volume of minority carriers within the base region. When the amount of base current is established by the external biasing circuitry, the transistor will contain a total minority charge within the base region that provides this amount of recombination or base current. In our two-dimensional model, Figure 3-3b, we can therefore think of the external base current as controlling the area under the minority-carrier concentration line (the cross-hatched area in the figure).

Reasons for Output Impedance: Early's Effect and Other Concepts

In 1952, Dr. James M. Early (then of Bell Laboratories) described the effects of base-width modulation, and this phenomenon is now named after him. As can be seen in Figure 3-4, when a fixed base-emitter ON voltage is applied to a transistor and the base-collector voltage is increased from a low value (V_{CBL}) to a higher value (V_{CBH}), the effective

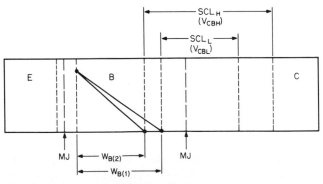

Fig. 3-4. A model of Early effect.

base width narrows from $W_{B(1)}$ to $W_{B(2)}$. This increases the slope of the minority-carrier concentration in the base region and therefore increases the collector current. This implies that a transistor has a finite output impedance; that is, changes in output voltage cause changes in output current. This is not desirable in an amplifying device because, ideally, output current should depend only on the input signal. The voltage gain of a transistor amplifier will be reduced because of this output impedance, because a portion of the output signal current ($g_m\ V_{in}$) will be diverted through this output impedance and is therefore not flowing in the external collector load resistor. This effect is more of a problem for large values of load impedance. For an example, the voltage gain is reduced by a factor of 2 when the output impedance of the transistor is equal to the external-stage load resistance. We will look at this in more detail shortly.

Another characteristic of transistors can be easily understood by using the simple model of Figure 3-4. Anyone who has operated a transistor curve tracer has observed the collector characteristics shown in Figure 3-5. Displays from these tracers permit determination of the dc character-

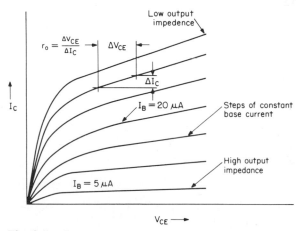

Fig. 3-5. Curve-tracer display of the collector characteristics of a transistor.

istics of a transistor. The method is to drive the base of a common-emitter–connected transistor with a sequence of fixed current levels. For each base-current step, the collector voltage is swept from 0 V to a maximum voltage (the collector voltage becomes the x axis) and the resulting collector current is displayed on the y axis. In this way, a family of curves is presented, one for each base-current step. At low values of base-current drive (say 5 μA), low collector currents result and this curve is "seen" to be horizontal. This implies that changes in the collector-base voltage are causing smaller changes in the collector current (although the percentage change in collector current is always the same), thus displaying a high output impedance. Notice that at large values of collector current, the output impedance is definitely much lower (because a given increase in V_{CE} will cause a larger increase in I_C).

This fact can be understood by looking at the transistor model in Figure 3-6. This figure shows the effects of base-width modulation on

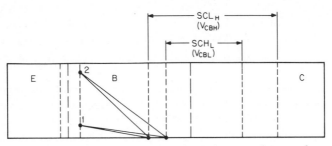

Fig. 3-6. Reduction in output impedance at large values of collector current.

two different operating-current levels: a low value of collector current, given by slope 1, and a high value of collector current, given by slope 2. As the effective base width changes, there is less of a change in slope for the low-current case, slope 1 (and therefore a smaller collector-current change, which corresponds to a larger output impedance).

This simple fact forms the basis of the curve traces shown in Figure 3-5 and the performance of typical transistors. A good estimate of the common-emitter output impedance r_0 of an IC NPN transistor is given by

$$r_0 \approx \frac{200}{I_C}$$

where it can be seen that low values of I_C provide high values of r_0. For an example, at $I_C = 1$ mA,

$$r_0 \approx \frac{200}{1 \times 10^{-3}} = 200 \text{ k}\Omega$$

A Simple Circuit Model for the Transistor

Now we can combine the preceding information to create a simple circuit model for the common-emitter transistor, as shown in Figure 3-7. This

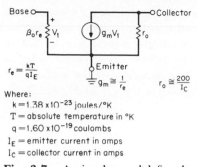

Where:

$k = 1.38 \times 10^{-23}$ joules/°K
T = absolute temperature in °K
$q = 1.60 \times 10^{-19}$ coulombs
I_E = emitter current in amps
I_C = collector current in amps

Fig. 3-7. A simple model for the monolithic NPN transistor. This model is only for dc and low frequencies.

model is a simplification of the popular hybrid-Π model and is restricted to direct current and low frequencies because all capacitors and other ac affects have been omitted. The model is still useful for circuit design, and a simple example will be given in the next section to illustrate the value of a transistor model in predicting the performance of a circuit.

Using the Transistor Model to Calculate Voltage Gain. As an example of how transistor models are used, a typical common-emitter amplifier circuit is shown in Figure 3-8*a*. For clarity, we have not shown the added biasing circuitry necessary to ensure that the transistor is conducting the 1 mA shown.

The 5-kΩ resistor R_C connected to the 10-V_{DC} power supply is called the *load resistor* for the transistor (we are assuming, for simplicity, that this is the only resistor connected to the collector). When we substitute

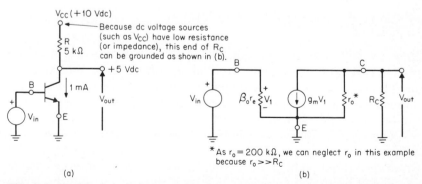

Fig. 3-8. Making use of the transistor model to calculate voltage gain: (*a*) the circuit (for clarity, the biasing circuitry is not shown); (*b*) modeling to calculate voltage gain.

the circuit model for the transistor, we obtain the circuit of Figure 3-8*b*. Note that the load resistor is small in value (5 kΩ) when compared with the much larger output impedance r_0 (200 kΩ) of the transistor. In this case, we can therefore simply omit r_0 because it will not have a significant effect on the accuracy of the gain calculation.

The resulting circuit is shown in Figure 3-9*a*. The output voltage

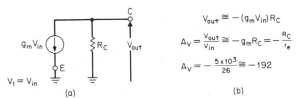

$$V_{out} \cong -(g_m V_{in}) R_C$$

$$A_V = \frac{V_{out}}{V_{in}} \cong -g_m R_C = -\frac{R_C}{r_e}$$

$$A_V = -\frac{5 \times 10^3}{26} \cong -192$$

(a)   (b)

Fig. 3-9. An example of the calculation of voltage gain: (*a*) the final circuit; (*b*) calculating the dc and low-frequency voltage gain.

V_{OUT} is produced when the current of the current source $g_m V_{in}$ flows through R_C. This provides the starting equation of Figure 3-9*b*, and when the numbers are substituted, we see that the gain is approximately (owing to all our simplifications) −192. The minus sign means that the gain stage is inverting (for example, as an input ac signal swings positive, the resulting output ac signal swings negative). Because we have neglected all capacitors and frequency dependence in this simple transistor model, the gain applies only to low frequencies or direct current. (The design of the dc biasing network and the proper accounting of the ac effects on voltage gain are the concern of the circuit-design engineer.)

Effects of Biasing Conditions on Output Impedance

Here we will consider the output impedance that results from three biasing conditions (1) base-emitter voltage drive, (2) base-current drive, and (3) emitter-current drive (Figure 3-10). The relative values of output impedance will be determined for each biasing condition by making use of the simple transistor model.

The idea of *voltage drive* from *voltage sources* and *current drive* from *current sources* may be new. The voltage source shown in Figure 3-10*a* is an idealization, which means that it will supply whatever current is demanded by the circuitry attached to it without causing the terminal voltage to drop in value. This is like a *good* battery in your car; such a battery keeps the headlights at essentially the same brightness (indicating

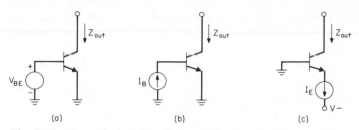

(a)   (b)   (c)

Fig. 3-10. Considering the effects of biasing conditions on output impedance: (a) common-emitter with voltage drive; (b) common emitter with current drive; (c) common-base with emitter current drive.

that the voltage is constant) even though you may be using the high-current-drain starter to crank the engine. Because the voltage from a *bad* battery will drop when supplying a large output current, the lights will dim while the starter is engaged. This is a simple test for battery trouble in your car.

Practical ideal voltage sources are provided by IC voltage-regulator circuits. Many of these have a source resistance (or output resistance) of only a few milliohms and are therefore close approximations to an ideal voltage source (which has a source resistance of 0 Ω).

Ideal current sources, such as those shown in Figure 3-10b and c, are harder to imagine. Suppose we want a constant current source of 1 mA. This means that this conceptual device will deliver 1 mA of current to a circuit node or a load resistor independent of the voltage that may develop at that node or across the load resistor. You can short-circuit a current source, and the 1 mA simply flows out one lead and back into the other. (If you short-circuit an ideal voltage source, *an infinite current* could flow. Enough current will actually flow to maintain this magnitude of voltage across the resistance of your shorting wire.)

Generally, current sources are realized by using an electronic circuit. For a *feeling* about current sources, consider the 1-mA current source shown in Figure 3-11. This current source drives the load resistor R_L. It will supply 1 mA of current as long as R_L is smaller than 10^9 Ω and

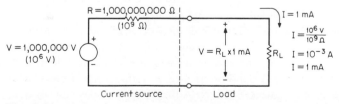

Fig. 3-11. A 1-mA constant-current source.

the resulting terminal voltage is smaller than 10^6 V. A few simple calculations using *reasonable* values for R_L will convince you. Now you see why a few transistors are used to make practical current sources. Even *real* current sources have an upper limit on the voltage they will provide, and this is called the *compliance voltage*.

The common-emitter with voltage drive (Figure 3-10a) was considered in our discussions of the Early effect. We have seen that this stage does have a finite output impedance, and we will use this as a reference for comparison of the relative output-impedance values for the connections in Figure 3-10b and c.

If we look again at Figure 3-10b, the case of base-current drive, we can set up the transistor model for this case (Figure 3-12). Remember

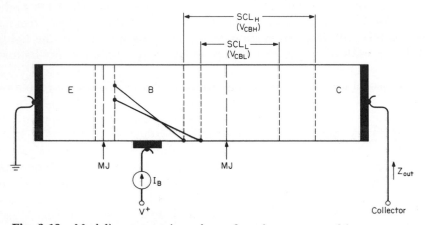

Fig. 3-12. Modeling output impedance for a base-current drive.

that constant base current forces a fixed amount of recombination in the base region. Therefore, a fixed volume of minority charge carriers will be established in the base region. Now, when the effective base width is narrowed (by increasing the collector-base reverse-biasing voltage), both ends of the line (representing minority-carrier concentration in the base region) must change. This causes a larger change in slope than the case of base-emitter voltage drive. Therefore, the condition of *base-current drive has a lower output impedance*.

To complete the comparison, now look at the case of emitter-current drive in the common-base connection of Figure 3-10c, as is modeled in Figure 3-13. The current generator in the emitter lead will constrain the emitter-injected current to be constant. This means that the slope of the minority-carrier concentration in the base region also must be constant. Now, as the collector-base voltage is changed, the slope must

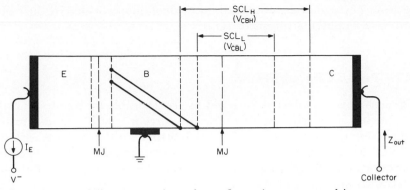

Fig. 3-13. Modeling output impedance for emitter-current drive.

stay constant. Both ends of the line are seen to move to keep the slope the same (since the emitter-base voltage does not have to be constant now). Consequently, this keeps the collector current constant and *provides the largest value of output impedance* for the three circuits considered in Figure 3-10.

The Cascode Connection

The feature of emitter-current drive is used in the *cascode* connection shown in Figure 3-14. A second transistor (Q_2) is added to the first

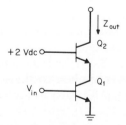

Fig. 3-14. The cascode connection.

transistor (Q_1) to provide a higher output impedance for Q_2. Therefore, a larger value of voltage gain can be obtained with this circuit. The added transistor (Q_2) makes use of a dc biasing voltage of 2 V to allow sufficient operating collector voltage for Q_1 (approximately $2 - 0.7$ V, or 1.3 V_{DC}). The collector of Q_1 provides essentially a constant current drive to the emitter of Q_2. The cascode connection also has other benefits in high-frequency circuits, since the output is now also isolated from the input because of the output-to-input shielding provided by the common-base amplifying transistor (Q_2).

3.2 SUPER-β TRANSISTORS

In general, narrow effective base widths are desirable for bipolar transistors. Such base widths not only raise the current gain β, but also result in higher-frequency transistors: it takes less time for the minority carriers to diffuse across a narrow base region (from the emitter to the collector). The disadvantages of a narrow base width are reductions in both the breakdown voltage and the output impedance of the transistor. The exchange between β and breakdown voltage is routinely made by varying the base width of the transistor during the fabrication process (as we will soon see, this control is obtained during the emitter-diffusion step).

A problem results if the base width is too narrow. As can be visualized, it could happen that with increases in the collector voltage, the SCL of the collector-base junction could sweep clear through the base region, jumping across the emitter-base SCL because no available charges exist there. This would cause the collector-base SCL to reach the emitter, where mobile carriers can be easily swept up, resulting in a large current flow from collector to emitter. This phenomenon is called *punch-through* and can occur in super-β transistors for collector-to-base voltages of only 3 to 5 V. Since such low-breakdown-voltage transistors are generally not useful, these punch-through-limited devices are discarded by the manufacturers of discrete transistors.

In the early days of linear ICs, Robert J. Widlar, a circuit-design engineer, noticed the extremely large values of β (3000 to 10,000) that could be obtained from these *defective* transistors. He used these transistors in the LM 108 op amp. Special circuitry had to be added to ensure that large voltages are never applied to these special *super-β transistors*. The LM 108 was a large step forward in the reduction of input current in monolithic op amps.

3.3 LARGE-SIGNAL EFFECTS

Most transistor models are limited to small-signal conditions. This means that the peak value of the input voltage must be small compared with V_T (26 mV). In addition, the mobile carriers must be in small concentrations when compared with the doping levels used in the various semiconductor regions.

As a transistor conducts larger and larger values of current, a point is reached at which the low-level, or small-signal, conditions are no longer valid and the transistor operation starts to be affected by new, large-signal mechanisms.

Kirk Effect or Base Stretching

Recall from earlier discussions that an SCL extends across the metallurgical junction to expose a certain amount of fixed charge in the lattice. When a transistor conducts current, mobile charges move through these SCLs. Under low-level, or small-signal, conditions, the density of mobile carriers is small compared with the impurity doping density (the density of uncovered fixed charge in the lattice). The presence of mobile charge within the SCLs is therefore neglected.

At high current levels, the mobile charge can no longer be neglected because its presence causes the SCLs to shift. An enlarged collector-base SCL is shown in Figure 3-15. This is for an NPN transistor, so

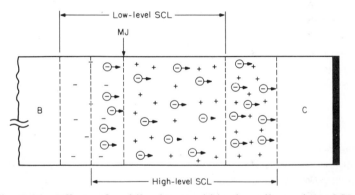

Fig. 3-15. Effects of mobile charge within the collector-base SCL.

the mobile charge carriers are electrons. The added mobile negative charge will affect the SCL on the two sides of the junction in different ways. On the base side, the added mobile negative charge means that the SCL does not have to extend as far into the base region to uncover the total negative charge needed. This increases the effective base width. This effect was described by C. T. Kirk, Jr. and is now known as the *Kirk effect* (or *base-width stretching*). The Kirk effect reduces both β and the frequency response of the transistor at large operating currents.

The negative mobile charge on the collector side cancels some of the fixed positive charge and therefore causes the edge of the SCL to move farther into the collector region in order to uncover more compensating positive charge. This has a minimal affect on transistor performance (it simply reduces the bulk resistance in the collector region) because the SCL moves closer to the collector contact.

Emitter Efficiency

Under large-signal conditions, the concentration of minority carriers at the emitter edge of the base region is considerably increased. There is a fundamental rule of semiconductors called *space-charge neutrality*. Simply stated, space-charge neutrality means that every small region within a semiconductor must contain a balance of opposite charges under unbiased or thermal-equilibrium conditions (it would take extremely large fields to cause charges to separate and remain separated and these large fields do not exist within the base region, although they do exist within the space-charge layers). One consequence of space-charge neutrality is that at large values of collector current, where the minority-carrier concentration has increased at the emitter edge of the base region, the majority-carrier concentration also must increase. Now we no longer have an N^+P^- diode, since the presence of this increased concentration of holes (majority carriers) acts to effectively create an N^+P^+ diode. More holes are therefore now back-injected into the emitter. This causes the base current to increase and β falls off at large values of collector current because of this loss of emitter efficiency.

Saturation

When bipolar transistors have an excessive base-current drive, the collector voltage can be driven down to very low levels ($V_{CE(sat)}$), as shown in Figure 3-16. Normal transistor action collapses (or saturates) because the collector-base diode now becomes forward-biased. The low value of V_{CE} when the transistor is in saturation results from the difference between the two forward-diode voltages ($V_{BE} - V_{BC}$). When a transistor

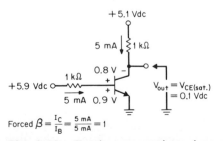

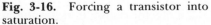

Fig. 3-16. Forcing a transistor into saturation.

is operated in saturation, the base volume is flooded with minority carriers and a large concentration of minority carriers exists in the collector region owing to heavy injection from the base. This is shown in Figure 3-17.

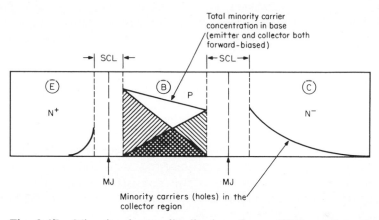

Fig. 3-17. Minority-charge distributions for an NPN transistor in saturation.

A transistor will not saturate if the external collector circuit can supply enough holes (current flow) to recombine with the electrons entering the collector region from the emitter. This demand for collector current becomes too large when it exceeds the limited current $I_{C(max)}$ that results because of a collector load resistor R_C, where $I_{C(max)} \simeq V_{CC}/R_C$ (neglecting $V_{CE(sat)}$). The *surplus electrons* that enter the collector by transistor action *cannot accumulate* in the collector region, *so they leave by causing the collector-base diode to become forward-biased.*

Saturation causes poor performance in high-speed switching circuits. For example, when the base current is switched to zero, it is desired that the transistor rapidly switch to an OFF state (as in T²L digital circuits, which allow the output transistor to saturate). Unfortunately, there is a time delay, or *recovery time,* during which the transistor stays ON, even though the input-control signal has changed to OFF. Saturated transistors are slow to turn OFF because of charge storage in both the base and collector regions; this is the reason for *gold-doping.*

Gold-Doping and Recombination

In the case of saturated logic, gold is diffused into the silicon lattice to purposely create recombination centers, or traps, near the center of the silicon band gap (as far as energy is concerned). These trapping centers capture both electrons and holes and provide a meeting place where they recombine and mutually annihilate each other.

Gold-doped switching transistors therefore more rapidly get rid of the minority carriers in the base and collector regions. Gold-doping is used in IC logic circuits, such as T²L, where the transistors are allowed to saturate. Unfortunately, the β of the transistor is also reduced when gold-doping is used. In addition, the gold ties up (compensates) some of the electrons available in the N-type collector region, thereby undesirably raising the resistance in the collector r_{sat}. As a result, V_{sat} increases at large values of collector current. Further, leakage currents also increase with gold-doping.

Not all logic families allow saturation. For example, high-speed emitter-coupled logic (ECL) is basically a differential amplifier with load resistors that are kept small enough in value to allow the transistors to remain active (out of saturation), even with the differential amplifier fully switched so that only one of the transistors is ON. This type of logic circuit is shown in Figure 3-18. Gold-doping is therefore not used in the manufacture of ECL circuits.

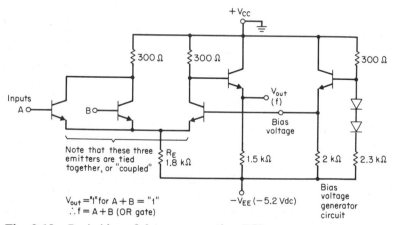

Fig. 3-18. Basic idea of the nonsaturating ECL gate.

The Baker Clamp

A number of years ago, Baker devised and published a circuit trick to prevent saturation. Baker's idea is shown in Figure 3-19. Essentially, two silicon diodes are added to steer excess base-drive current into the collector lead of the transistor. Now the collector voltage is *clamped* at approximately 0.7 V. As a result, the transistor cannot saturate because the collector-base junction cannot become forward-biased. In addition,

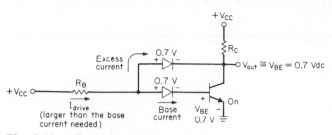

Fig. 3-19. The Baker clamp to prevent saturation.

the clamp means that the collector voltage no longer has the low $V_{CE(sat)}$ of standard saturating T²L.

A Schottky Diode Clamp

Today, a metal-semiconductor diode, or Schottky diode, is used on both Schottky (S) and low-power Schottky (LS) logic families to similarily prevent saturation (Figure 3-20). The forward voltage of the Schottky

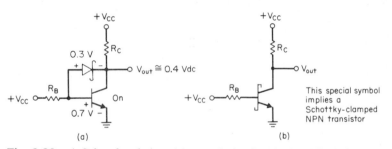

Fig. 3-20. A Schottky clamp: (*a*) actual circuit; (*b*) simplified drawing.

diode is approximately 0.3 V. Once again, the output voltage does not swing as low as with standard T²L. Low-power Schottky provides essentially the same speed as T²L, but with much less power drain. It therefore has become a very popular logic family.

CHAPTER IV

How Transistors Are Made: A Short History of Manufacturing Technologies

The key problem with transistor fabrication is to find a way to make an extremely thin base width—and to do it with good control. Early techniques did not employ diffusion furnaces or ion-implanting machines. In fact, when the transistor was discovered by J. Bardeen and W. H. Brattain in 1947 at Bell Laboratories, two closely spaced metal needles were simply brought in contact with the same germanium "base" wafer to which an ohmic contact was made. One needle served as the emitter and the other as the collector of this *point-contact transistor*. The name *base* for the third electrode, the semiconductor wafer, was carried on, and all the names originally used for transistor leads are still in use today. Point-contact transistors were hard to produce commercially, so other ways to employ this new transistor concept had to be developed.

William Schockley, also of Bell Laboratories, theoretically described PN junction transistors in 1949. His idea was to form a three-region structure (NPN or PNP) within a common semiconductor material that has two parallel junctions very near each other. In this case, the relatively thin center region is the *base* and the end regions are the *emitter* and the *collector*. This construction technique brings the collector region close enough to the emitter region to allow the collector to steal essentially all of the emitted current. This is *transistor action*. The first practical junction transistor was the germanium-alloy transistor.

4.1 GERMANIUM-ALLOY TRANSISTORS

The basic approach to producing early germanium-alloy transistors was to grow a single crystal of N-type germanium and then to slice this up into relatively thin wafers (approximately 20 milli-inches thick). Two small pellets of indium (a P-type dopant) were then placed on the resulting "pills," or small wafers of germanium. A large-diameter indium pellet was placed on one side of the wafer, and this became the collector. A smaller-diameter indium pellet was placed on the other side, and this became the emitter. The complete structure is shown in Figure 4-1a.

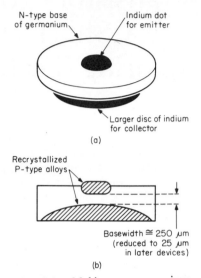

(a)

(b)

Fig. 4-1. Making a germanium-alloy transistor: (a) piece parts in position; (b) cross section after alloying.

This three-piece assembly was heated in a high-temperature furnace, which caused the indium to alloy with the germanium. During the cooling cycle, the P-type regions of germanium-indium alloy recrystallized, forming the emitter and the collector. The result was a PNP transistor.

Base-Width Control

If the thickness of the starting wafer of germanium varied, changes in the finished base width of the transistor resulted. As mentioned, base-width control was generally poor with this alloy process. As a result, many of the transistors had relatively wide base widths (up to 250 μm). These transistors had better forward gain than reverse gain (measured with the emitter and collector leads interchanged) simply because current injected from the small pellet was more likely to be collected by the large pellet. When the roles of emitter and collector were interchanged, to measure inverse β, much of the current now injected by the large pellet simply was not collected by the small pellet and therefore recombined within the base region and reduced β. Since the emitter and the collector were of the same doping concentration, the emitter-base breakdown voltage was equal to the collector-base breakdown voltage (this is no longer true in modern transistors).

The excessively wide base widths therefore caused low β and poor high-frequency performance. A technique was needed that did not depend on the initial thickness of the wafer and the poorly controlled alloying process to establish the base width. Many ingenious techniques were attempted (and many management teams were consumed) before researchers considered building a transistor that involved operations performed on only one surface. This resulted in the birth of the *planar* transistor in the late 1950s.

4.2 DISCRETE-PLANAR, EPITAXIAL, PASSIVATED-SILICON TRANSISTORS

The lower leakage currents and wider operating temperatures of silicon transistors, as compared with germanium transistors, stimulated continued efforts to bring these devices to the marketplace. Prior to the appearance of silicon transistors, many complex fabrication techniques had been used to provide relatively high-frequency germanium transistors. These products presented strong competition for the early silicon transistors—especially the lower-performance silicon-alloy transistors. However, the shift to silicon changed many of the ground rules and, in fact, provided many fabrication advantages. Let us find out why.

Oxide Benefits with Silicon

Since the oxide of germanium will not block the diffusion dopants (that is, it will not stop the diffusion of dopants into the germanium), a different material was necessary for use in the planar transistor. That material was silicon. The oxide of silicon, SiO_2, not only masks the impurity dopants, it also provides an excellent protective cover for the surface of the device and is an excellent insulator. Amorphous SiO_2 forms easily on silicon wafers when they are heated in an oxygen atmosphere. To facilitate the process, this oxidation is done in tubes made of quartz, which is the crystalline form of SiO_2. The benefits of having SiO_2 on the surface of the wafers and the planar process were major reasons for the rapid progress of silicon transistors in the 1960s, which followed the decade of germanium transistors. The next subsections will explain how and why this occurred.

Epitaxial Layers

Several technological refinements aided the development of planar silicon transistors. One such refinement was the use of an epitaxial crystal-growth technique to provide an excellent collector region for the transistor. *Epitaxial crystal growth* is a way that new material can be added to, or grown upon, an existing crystal. This new material, in the case of epitaxial silicon deposition, is usually obtained from the thermal decomposition of a gas that contains silicon, such as silane (SiH_4) or silicon tetrachloride ($SiCl_4$). This deposition takes place within a special piece of processing equipment called an *epi reactor*. In addition, impurity dopants also can be entered into the gas flow to produce an epitaxial layer with a given dopant type (N or P) and resistivity. The new growth replicates the crystal structure of the starting wafer.

With epitaxial crystal-growth techniques, transistor manufacturers could now abruptly change doping concentrations and control thickness of the starting surface, as shown in Figure 4-2. The process works as

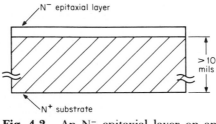

N^- epitaxial layer

> 10 mils

N^+ substrate

Fig. 4-2. An N^- epitaxial layer on an N^+ substrate.

follows. A richly doped (to reduce r_{sat}) N^+ substrate is made thick enough to minimize breakage in handling. Onto this substrate, a *controlled* thickness of N^- (to obtain high breakdown voltage) silicon is grown by epitaxy (such a uniformly doped layer is impossible to obtain with diffusion processes). Then, diffusion furnaces, operating at temperatures of 900 to 1200°C, are used to add impurity dopants into selected areas of the thin epitaxial surface layer. These dopants diffuse into the silicon in much the same way carbon diffuses into iron to form steel.

Oxide Masking

The key to the planar process, however, is the dopant masking naturally provided by the SiO_2 layer. The three possible P-type dopants for silicon are boron, aluminum, and gallium, but only boron can be masked by SiO_2. Consequently, boron is used for all P-type regions in most silicon products. The N-type dopants for silicon are antimony, arsenic, and phosphorus, and they all can be masked by SiO_2. So let us examine how masking is done.

After epi growth, the wafer is oxidized. This encases the wafer in a protective oxide shell. Selected areas of this oxide layer are then removed by a wet-chemistry etchant by making use of a precise photolithographic masking process.

To provide the masking, a wafer is first coated with a special light-sensitive chemical called *photoresist*. A glass plate with a desired pattern in an opaque emulsion, the mask, is then placed over the wafer, and ultraviolet light is allowed to pass through the clear areas of the pattern. This causes a change in the photoresist film on the wafer. The light induces cross-linking of the photoresist molecules, which hardens the photoresist layer (this is called *negative resist; positive resist*, which is hardened by the absence of light, is also available). Those parts of the resist film not exposed to light are later dissolved away in a developing operation. What is left is a controlled pattern of hardened photoresist. This photoresist pattern serves as a varnish that allows only selected areas of the oxide surface layer to be removed in a subsequent wet-chemistry SiO_2 etching step.

The result of this photomasking and etching process is a silicon wafer on which only specific areas of the surface (the mask pattern) are exposed. As a result, when the wafer is put into a P-type diffusion furnace, only the selected areas are doped (by solid-state diffusion of the dopant into the silicon crystal). The results of these processes are diagramed in Figure 4-3.

Notice that the dopant also diffuses laterally under the SiO_2 layer. This was a major breakthrough in controlling the surface conditions of semiconductors, since the surface junction could be protected by SiO_2. This resulted in greatly increased reliability and improved electrical

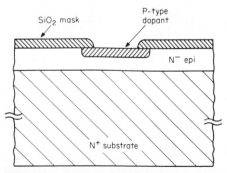

Fig. 4-3. Selectively doping to produce the base regions.

characteristics for the products. So far in our examination of the planar transistor production process we have an N⁻ collector region (the N⁻ epi) and a P-type base region. All we need now is an N⁺ emitter to complete the NPN transistor structure.

As the P-type dopant is diffused into the crystal, an oxidizing atmosphere is set up in the furnace so that a new layer of SiO_2 will cover up, or cap, the surface of the newly formed P-type base region. This new SiO_2 layer is now in place and ready for the next masking step. The N-type dopant for the emitter is now patterned within the new oxide layer by another photomasking and etching process. As before, this allows the dopant to enter the silicon crystal only in selected areas inside the previously diffused base region. During the emitter diffusion, moreover, the surface is again oxidized for protection. Now we have a completely formed NPN transistor under an SiO_2 passivation layer.

The base width for such a transistor is determined by the difference in the depths of the emitter and base diffusions. With this process, this difference in depth can be kept both thin (0.5 to 3 μm) and under relatively good control. This is the key to the high-frequency performance of the planar transistor.

Undesired Emitter Pipes

Narrow base widths do present problems in the manufacture of transistors, however. Crystal-lattice defects cause very small regions in which the emitter dopant can selectively diffuse into the silicon at a faster rate. These small *pipes* of emitter material can even extend from the emitter clear through the base region into the collector. This places

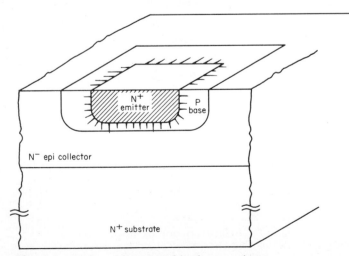

Fig. 4-4. Emitter pipes in a bipolar transistor.

conducting channels from the emitter to the collector which are, in fact, submicroscopic, circular, junction field-effect transistors, as shown in Figure 4-4. If the cross section of these pipes is small enough, the current that passes through them will limit or saturate as the collector voltage of the transistor is increased (this JFET action will be discussed in Chapter 6). Some of these pipes may not actually touch the collector region, but as the collector-base SCL extends further into the base region with increases in collector voltage, such shallow pipes can be uncovered. Special precautions are therefore used in wafer fabrication to keep this problem at acceptable levels, but the defect still tends to reduce yields on many large-die-area bipolar IC products. [It is also thought that these pipes cause mismatch in V_{BE} (causing offset voltage in linear products) and may be responsible for the excessive noise commonly called "popcorn" noise.]

Providing Contacts to the Transistor

Photomasking is again employed to expose small areas of the emitter and base regions so that a deposited layer of aluminum can be used to make ohmic contact with these doped regions of the transistor.

At first, vacuum-deposited aluminum metal completely covers the wafer. It is then photomasked and mostly etched away. Only aluminum bonding pads (for the eventual connecting of small bonding wires for the base and emitter contacts) and short metal runs from these pads over to the base and emitter contacting regions are left on the wafer surface.

A contact for the collector is not needed at this point because it is supplied when the transistor chip is gold brazed or soldered to a metal header, the transistor package. This is why silicon transistors in metal cans (TO-3, TO-5, and TO-18, for example) have a *hot* can—the container is both electrically and physically tied to the collector.

This manufacturing technology is still in use to build discrete transistors. Today many thousands of small-signal transistors are built on one large wafer. After wafer fabrication is complete, each transistor is electrically tested and the defective units are ink-marked before the wafer is scribed and broken up into individual transistor chips. The uninked chips are then mounted, one to a container, and two separate bonding wires are used to attach the bonding pads to the base and emitter pins of the can.

A Completed Transistor

A cross section of a completed transistor chip is shown in Figure 4-5. Notice that there is no contact for the collector (this will be provided

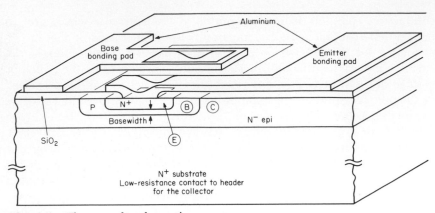

Fig. 4-5. The completed transistor.

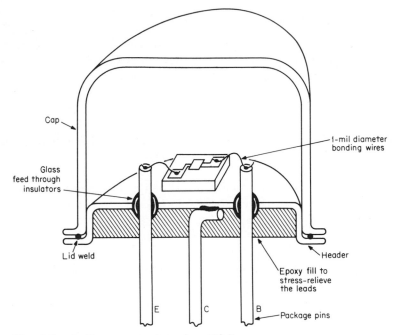

Fig. 4-6. A discrete transistor in a T0-5 can.

when the chip is mounted in the can). Figure 4-6 shows a transistor die-bonded to a metal header. Only two wires are needed, and only two pins feed through the header; the collector lead is butt-welded to the outside of the header.

CHAPTER V

Bipolar Integrated Circuits: The Whole Thing on One Chip

The standard bipolar monolithic (which literally means "one stone" and is used to imply a single, solid structure) circuit involves processing that is very similar to that of the discrete transistor. However, the process had to be changed somewhat in order to get multiple, isolated transistors on the same chip. In addition, the resistors which are needed in the circuits can be obtained from the planar diffusions which are used to make transistors. Consequently, the development of the IC (in 1958) came on the heels of the development of the silicon planar transistor.

5.1 SOME BASIC FACTORS

The building of complete circuit functions on a single silicon chip when only NPN transistors and resistors were available created many problems for the early monolithic circuit designers. It is interesting to see how unusual transistors with special geometries as well as new concepts in circuit design have evolved to solve all these problems.

Linear versus Digital Circuits

Use of the term *linear circuits* separates these circuits from the ON, OFF, or *digital logic,* circuits. ON/OFF circuits, or *binary (two-state) logic circuits,* generally have large input-voltage swings (either 0 V or 5 V), and the output voltages of these circuits are also either 0 V or 5 V. In contrast, linear circuits typically operate with low-level input-voltage swings (sometimes only 1 μV or less, such as in a radio receiver). Similarly, the output voltage of a linear circuit is represented by relatively small deviations about a "bias" (or "quiescent," which means "quiet") dc operating point. Therefore, digital circuits have some "noise immunity" (typically 400 mV), and linear circuits have no noise immunity. Linear circuits therefore create much more trouble if operated in an electrically noisy environment. Further, digital circuits are less dependent on the detailed performance characteristics of the circuit components and will usually either continue to function properly or completely fail as the ambient temperature is changed (as opposed to a linear system, which may drift out of specifications). Finally, in contrast with linear ICs, digital ICs usually can be interconnected directly, without use of external (and sometimes critically valued) resistors or capacitors (except for the relatively noncritical power-supply bypass capacitors which, like linear ICs, are still needed, especially for memory ICs).

Moreover, the continual decline in the costs of digital ICs is encouraging conversion of previously analog electronic systems to digital systems. As a result, there has been a tremendous increase in total transistor count. Thus we are finding relatively sophisticated digital computers (microprocessors) performing many relatively simple tasks, because "computers are free" (or nearly so).

Isolating the Transistors

Many individual transistors are necessary to build complete circuits on a single semiconductor chip. To obtain multiple, electrically isolated transistors, the doping of the starting substrate is changed to P⁻. Further, as a way to help reduce bulk resistance in the collector regions, a buried N^+ layer is selectively diffused into the P⁻ substrate prior to addition of the N⁻ epitaxial layer. These changes result in the initial cross section shown in Figure 5-1.

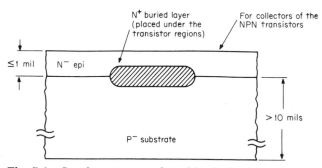

Fig. 5-1. Starting structure for a bipolar integrated circuit.

Following this, separate N⁻ epi regions, or "tubs," must be formed to isolate the multiple NPN transistors. This is done by using a deep P^+ diffusion that isolates these many epi tubs from each other (Figure 5-2). This is called *diode isolation,* because all the epi-substrate diodes are always reverse-biased. This is the key to building complete circuits on one chip.

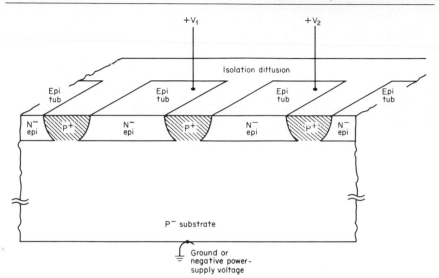

Fig. 5-2. Junction-diode isolation of N⁻ epi tubs.

Fabricating NPN Transistors

NPN transistors are made in essentially the same way as discrete transistors are made: a P-type base diffusion is located within an isolated N⁻ epi tub. The differences are few. One is that the emitter diffusion is also allowed to enter the N⁻ tubs in the areas where the metal contact for the collector will eventually be made. This change was necessary because the N⁺ substrate is no longer available as the collector contact. The collectors therefore have to be contacted on the surface of the die. (Use of a relatively high-concentration emitter diffusion prevents the formation of a Schottky diode, which occurs if the aluminum interconnect metal is placed directly on the lightly doped epi region.)

To help reduce the resistance of the N⁻ collector-bulk regions, an N⁺ buried layer is used, as shown in Figure 5-3. The emitter current crosses the narrow base width of the transistor and then passes through only a thin region of the high-resistivity N⁻ epi region (shown as region 1) before it enters the low-resistivity N⁺ buried layer (region 2). Another thin N⁻ epi region (shown as region 3) then has to be crossed before the collector current finally reaches the N⁺ surface-collector contact diffusion. The presence of this N⁺ buried layer reduces the series resistance of the collector from thousands of ohms to a few hundred ohms in the standard small-signal (small-geometry) transistor. There is still a larger collector-bulk resistance problem with IC transistors than with discrete transistors. This undesirably raises the collector saturation voltage of the IC devices.

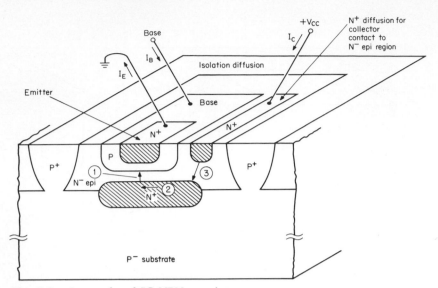

Fig. 5-3. A completed IC NPN transistor structure.

The P-type base diffusion is generally also used to make the resistors needed in most circuits. Typically, these diffused resistors are placed in an isolated N^- epi tub and this tub is tied to the most positive potential of the circuit to always keep the PN (resistor-to-epi) diode reverse-biased.

Notice that more processing steps and complications are introduced in this process as compared with discrete-transistor fabrication. For power ICs, a special deep collector N^+ contacting diffusion is often used, and this surface diffusion actually is driven into the buried layer. This reduces the vertical component of r_{sat} under the collector contact (which is shown as region 3 in Figure 5-3). In operation, the heavy emitter-current flow rapidly conductivity modulates (or reduces the apparent resistivity owing to the presence of charge carriers) the resistance of the collector region, which is directly under the emitter (region 1 in Figure 5-3).

The Current Mirror and ΔV_{BE} versus ΔI_C Relationships

A trick of monolithic circuit design is to voltage-bias the base-emitter junctions of a group of transistors operating in parallel. This approach will not work with discrete transistors because of large differences in the V_{BE} versus I_C relationships of individual devices.

From the ideal-diode equation we can determine the increase in diode voltage necessary to cause the diode current (or the collector current in a transistor) to double as well as to increase by a factor of 10. This can be calculated by considering two forward currents, I_1 and I_2, that can be expressed as

$$I_1 = I_S \ (\exp \ V_1/V_T)$$

and

$$I_2 = I_S \ (\exp \ V_2/V_T)$$

where:

$$V_T = \frac{kT}{q}$$

Now if we take the ratio of I_2 to I_1, we obtain

$$\frac{I_2}{I_1} = \frac{I_S \ (\exp \ V_2/V_T)}{I_S \ (\exp \ V_1/V_T)} = \exp \frac{V_2 - V_1}{V_T}$$

So, for a ratio of 2:1,

$$\frac{I_2}{I_1} = 2 = \exp \ \Delta V/V_T$$

where:

$$\Delta V = V_2 - V_1$$

or

$$\frac{\Delta V}{V_T} = \ln 2$$

$$\Delta V = V_T \ln 2 = (26 \times 10^{-3})(0.693)$$
$$= 18 \ \text{mV}$$

Similarly, for a 10:1 change in current,

$$\Delta V = V_T \ln 10 = (26 \times 10^{-3})(2.3)$$
$$= 60 \ \text{mV}$$

This means that to change the current of a diode (or a transistor, since the same equation is used) from its present value by a factor of 2 or ½ requires only a ±18-mV change in voltage; morever, ±60 mV gives a 10-fold change. This helps show why poorly matched discrete diodes or transistors cannot be simply voltage-biased. It is also useful to keep this in mind when troubleshooting a circuit, since precise measurements of transistor V_{BE} can be used to indicate the approximate value of the collector current flowing.

Current mirrors are used in biasing most linear monolithic ICs; they are also used as differential-to-single-ended converters in most monolithic operational amplifiers (op amps); and they have also been used to provide a noninverting input for the unusual Norton amplifier, or current-differencing amplifier, the LM3900. In addition, they have been used to overcome bias-circuit design restrictions as well as simplify the design of dc-coupled stages, which result because coupling and bypass capacitors are generally not available in the design of linear ICs.

If we have matched IC components on the same chip, we can see how the circuit of a current mirror works (Figure 5-4). The first thing

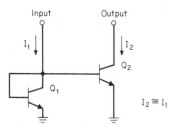

Fig. 5-4. The current mirror of IC designs.

to notice is that transistor Q_1 has a collector-to-base short. It is still acting like a transistor, and this is one way to force Q_1 to carry the input current I_1 as collector current (for simplicity, we will neglect the base currents of Q_1 and Q_2). We have Q_1 acting as a shunt regulator holding the V_{CE} at V_{BE}, so it is often said to be a *diode-connected transistor.*

Now a V_{BE} automatically results, and this allows the collector of Q_1 to carry I_1. This same V_{BE} is applied to the second (and well-matched) transistor Q_2. Thus Q_2 is biased also to carry the same amount of collector current. To a first order, $I_2 = I_1$, and this is the idea of the current mirror. The current I_1 is therefore said to be *mirrored* about ground because Q_2 now carries this value of current in its collector.

Emitter-Area Scaling

The transistors of the preceding current mirror could be designed to produce a larger value of I_2, so, for example, I_2 could equal $3 \times I_1$. This is done by making transistor Q_2 have an emitter area that is 3 times larger than that of Q_1 (or use three transistors in parallel). Increasing the emitter area causes I_S to be increased by the same factor (as compared with I_S for Q_1), such that for the same V_{BE}, a larger collector current will flow. This can be seen from the ideal-diode equation:

$$I = I_S \left[(\exp V_D/V_T) - 1 \right]$$

This fact is used in the design of most linear IC products, especially in dc biasing circuitry.

Available Compatible PNP Transistors in Linear ICs

In addition to the relatively high-frequency, high-β NPN transistor, two useful lower-frequency, lower-β PNP transistors also can be fabricated using the same processing steps. One of these transistors uses a separate N^- epi tub for the base region, and both the emitter and collector regions are formed by the same P-type base-diffusion process used in the standard NPN transistor. This results in transistor action that is lateral and no longer vertical, so the device has been called a *lateral PNP*. As expected, these devices have useful β only if fabricated within a relatively lightly doped N^- epi region, because the epi region becomes the base of the transistor. This has kept PNP transistors out of standard digital IC circuits (such as T^2L) because the epi is generally too heavily doped in these low-voltage circuits. However, PNP input transistors are used in many newer bipolar-logic products.

The details of lateral PNP structure are shown in Figure 5-5. The

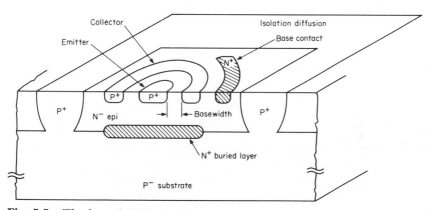

Fig. 5-5. The lateral PNP transistor.

N^+ buried layer is used to raise β. It may at first appear that this heavily doped N^+ region would increase the recombination of minority holes in the base region and thereby reduce β. The interesting thing is that an electric field naturally forms within the buried-layer diffused region. This occurs because the fixed charge in the buried layer is not uniform. An electric field results which repels the mobile holes and therefore keeps them out of the buried-layer region; therefore, β is increased.

Once minority carriers exist in a given region, they will not move into regions of higher doping concentration to be recombined because of this built-in repelling electric field. Therefore, undesired holes that may exist in the N⁻ epi region of an IC, for example, will not recombine in an N⁺ emitter diffusion that may be driven into the epi in an attempt to cause recombination. To get rid of the holes, a P-type diffusion should be used instead, since it can be biased and operated as a collector.

The base width of the lateral PNP is set by both the base photoresist step and the lateral diffusions of the P regions under the SiO_2 masking layer. Obviously, control of base width is much more difficult than in NPN devices, where base width is obtained by the difference in diffusion depths of the base and emitter regions. As a result, the lateral PNP transistor has a reduced high-frequency response as compared with the higher-performance NPN transistor. Further, β is lower and also rapidly decreases at high collector-current densities. The lateral PNP is still a very useful transistor for dc biasing applications, and many low-frequency circuits have made full use of this device.

Current Scaling in Multicollector PNPs

The lateral PNP easily allows multiple collectors to be obtained by simply segmenting a large circular collector into smaller, separate sections, as shown in Figure 5-6. In addition, the base width of each segment can be changed in the design of the transistor to give an additional control on the split of total emitter current among the multiple collectors. This built-in base-width variation causes a fixed, or built-in, Early effect and is very useful in achieving a wide range of separate dc biasing currents from a single multiple-collector PNP transistor geometry.

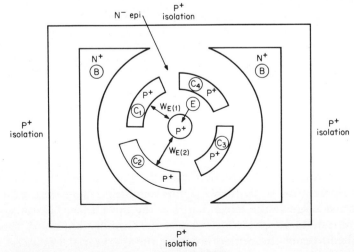

Fig. 5-6.   A multicollector lateral PNP.

When multiple-collector segments are used, the output impedance of these sectored collectors is lowered because changes in the base-collector SCL also affect the sector size, or capture angle, presented to the emitter, as shown in Figure 5-7. To raise output impedance and

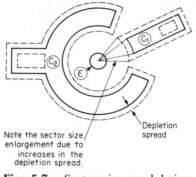

Note the sector size
enlargement due to
increases in the
depletion spread.

Depletion
spread

Fig. 5-7. Sector-size modulation. The percentage effect is worse for small sectors.

still take advantage of base-width scaling, multiple single collector PNP transistors should be used. These can be driven with their base-emitter junctions connected in parallel, and the base widths of the individual transistors can still be altered to provide current scaling.

Current Reduction with Reinjector PNPs

Lateral PNPs are useful devices. In one unusual application, a lateral PNP transistor can provide a small bias current for a circuit without requiring large-valued resistors. A reinjector PNP current divider is shown in Figure 5-8. This is a useful application of transistor theory. The idea is to start with an easily obtained current I_B and then reduce it to a much smaller value of current I_{out}, which can be a small fraction (as small as $\frac{1}{5000}$) of the I_B reference. The basis of this scheme is that a floating collector (such as C_2 of Q_1) will collect the holes provided by the emitter (E_1 in this example). These collected holes cannot accumulate in C_2, so they are *reinjected;* that is, C_2 saturates and turns ON *to simultaneously act as an emitter* to get rid of the collected holes. Notice that we have tied C_2 to E_2 of Q_2. This causes E_2 to participate in this reinjection, so it behaves as it should: it is the emitter of Q_2.

To obtain the desired current reduction, C_2 is made smaller than C_1, and this large ratio is maintained in Q_2 and Q_3. A current reduction of $\frac{1}{17}$ in each device provides the overall 1 : 5000 ratio. (This technique was used to generate a 2-nA reference current from an easier-to-obtain 10-μA reference in a custom IC design.)

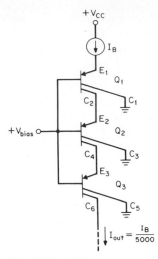

Fig. 5-8. Bias-current reduction using reinjector PNPs.

The layout sketch in Figure 5-9 shows that base-width scaling is also used in reinjector PNPs (C_2 has a wider base width than C_1). Collector C_2 of Q_1 is connected to emitter E_2 of Q_2. Conceptually, each of these transistors could be placed in one large N^- epi tub, because the bases of all the transistors are common. This was not done in favor of the larger degree of isolation provided by separate epi tubs for each. For example, in Figure 5-8 a few of the holes emitted by E_1 (of Q_1) could somehow be collected by C_6 (of Q_3) and could therefore cause a large error in the current division. The advantages of the lateral PNP device also have been exploited to create the integrated injection-logic circuits, and we will discuss these later in this chapter.

The Vertical PNP Transistor

Another useful PNP transistor can be realized in the standard linear IC process by simply omitting the N^+ buried layer and the surface P-type collector ring and allowing the P-type emitter to inject and both the P-type substrate and the P^+ isolation-diffusion sidewalls to act as collectors. In this way, the collector is tied to ground (or $-V_{EE}$). However, such a *vertical PNP transistor* does provide an improvement over lateral PNPs for emitter-follower applications. Actually, higher performance is obtained if the additional lateral-surface P-type collector is kept in place and is simply extended into the P^+ isolation region (to supply the ground contact), as shown in Figure 5-10.

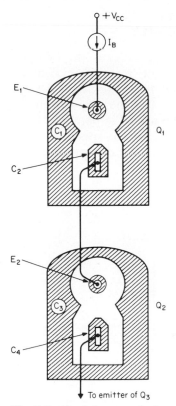

Fig. 5-9. Basic layout of rein-jector PNPs.

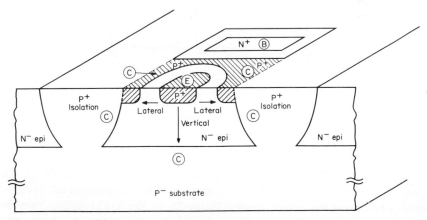

Fig. 5-10. Combining lateral collection with the vertical PNP.

5.2 T²L, THE POPULAR
BIPOLAR-LOGIC FAMILY

Transistor-transistor logic (T²L) is the final stage in the evolution of bipolar-logic circuits. The monolithic digital world worked its way through direct-coupled transistor logic (DCTL) to resistor-transistor logic (RTL). This eliminated the base-current hogging of one transistor by allowing many to be placed in parallel. Diode-transistor logic (DTL) was then introduced to raise the speed. These logic circuits are shown in Figure 5-11. Finally, all the diodes were placed in one multiemitter

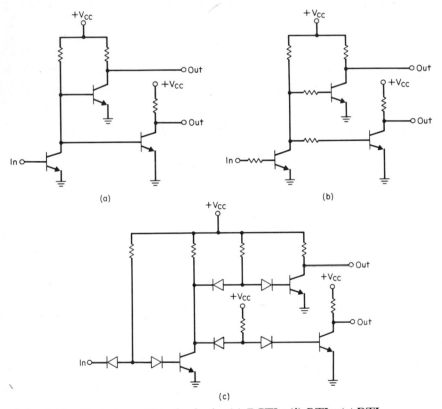

Fig. 5-11. Evolution of bipolar logic: (*a*) DCTL; (*b*) RTL; (*c*) DTL.

structure, resulting in the birth of T²L (Figure 5-12). The collector-base diode D_5 of the T²L input transistor replaced the diode at the base of the transistor in the DTL circuit and the multiple emitter-base diodes replaced the input diodes D_1 to D_4 in the DTL stage. The die area was therefore reduced, and in addition, the *transistor action* improved

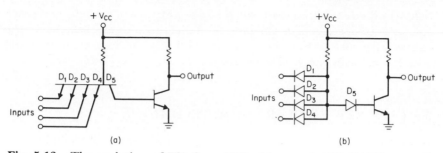

Fig. 5-12. The evolution of T²L from DTL: (*a*) an early T²L NAND gate; (*b*) the earlier DTL gate.

speed. When the input transistor turns ON (all input voltages low), the collector will extract charge out of the base of the output transistor and thereby help to reduce the ON-to-OFF transition. In addition, when the input transistor turns OFF (all input voltages high), the base region already has the required charge to allow the base-collector diode to conduct and thereby more rapidly turn the output transistor ON.

Special geometries (Figure 5-13) are used for the multiemitter input

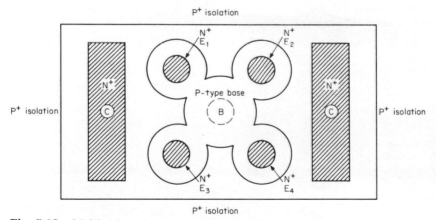

Fig. 5-13. Multiemitter input transistor of T²L.

transistor to reduce the lateral bipolar parasitic NPN transistor action that exists between the multiemitters used as inputs. This parasitic transistor action increases the current that must be supplied to an input that is held high while the other inputs are all simultaneously held low.

The T²L family is still available today, but it is being replaced by the newer Schottky-logic family, which is an evolutionary improvement of the same basic T²L circuit approach. Schottky logic makes use of a

Schottky diode to shunt the collector-base junction of the output transistor, as was described in Chapter 3. A Schottky diode will divert the excess base current into the collector, and this prevents saturation and greatly improves the switching speed.

5.3 I²L, A NEW BIPOLAR-LOGIC TECHNIQUE

Integrated injection logic (I²L) is an interesting way to get higher logic density on a bipolar chip. In addition, it has the more significant advantage of allowing linear circuitry (especially radio-frequency, power, or high-voltage circuits) to be combined on the same die. This is therefore an attractive choice for complex systems that require both linear and digital circuits on the same chip.

This relatively new logic family is unusual in that it has a single input lead and multiple outputs. These outputs are provided by a special multiple-collector NPN transistor geometry. To obtain multiple isolated collectors, the normal IC NPN transistor is operated upside down, in the inverted mode (this causes high inverse β to be an important design factor for these devices). The N⁻ epi tub now becomes the emitter, and separate N⁺ diffusions (this is the old emitter diffusion) into the P-type base region form the isolated multiple collectors, as shown in Figure 5-14. The drive current to the bases of these multiple-collector transistors is provided by a single P-type emitter of a very large lateral PNP transistor geometry. The individual base regions of the inverted NPN logic transistors form multiple collectors for this spread-out

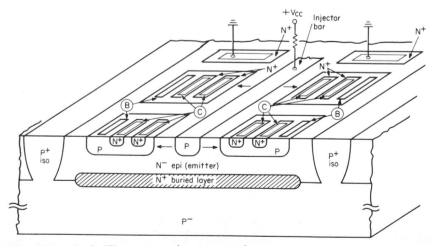

Fig. 5-14. Basic I²L structure in cross section.

P-type emitter, or *injector bar*. The circuit equivalent of this logic family is shown in Figure 5-15. Any collector of another logic gate can tie to the input of Q_2 and can clamp this base to ground. This will turn Q_2

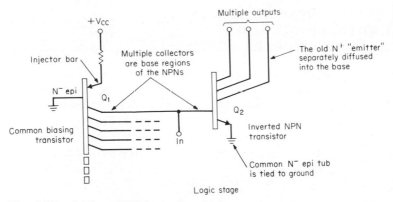

Fig. 5-15. Basics of I²L-logic circuits.

OFF. This structure, a single input with multiple outputs, causes the logic circuits to appear strange, as can be seen in Figure 5-16.

Ideally, all the logic can be laid out in one large N⁻ epi tub, and this saves area, because separate P⁺ isolation diffusions are not needed. Unfortunately, undesired interaction usually requires the use of additional P-type isolation regions or N⁺ diffusions that are deep enough to contact to the N⁺ buried layer (*N⁺ sinkers*), and these increase the total die area.

The main application of I²L is in circuits that combine logic with linear circuitry on the same chip. I²L has not been too successful as a

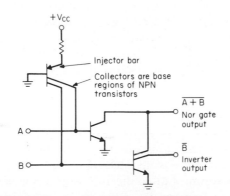

Fig. 5-16. An I²L NOR gate and inverter.

logic family, and the new advances made with MOS products in the area of high-density N-channel logic have precluded its widespread use.

5.4 DETRIMENTAL EFFECTS OF PARASITIC DIODES AND TRANSISTORS IN IC PRODUCTS

Unexpected parasitic diodes, transistors, and even SCRs can appear when input or output leads are forced either higher in voltage than the positive power supply or lower than the negative power supply (or ground, if a negative power supply is not used). This strange operation can usually be traced to a diode on the chip that has become forward-biased. When this occurs, the diode acts like an emitter for other *parasitic* transistors on the chip.

Epi Tubs Cause Problems

Another common problem occurs when an epi tub is made available to the outside world via a package pin. Such an epi tub could be the collector region of an NPN transistor, such as, for example, when this collector is tied to the output pin. In addition, the N-type base region of a PNP transistor, which is brought out of the package as an input pin, could allow external access to an epi tub. To add to these problems, many linear ICs use PNP input transitors, and newer digital-logic circuits are reducing input currents by also using PNP input transistors.

If we consider single positive-power-supply applications, it would appear that there is no way to obtain a negative voltage in the system that could cause these N-type epi tubs to be forward-biased with respect to the P-type substrate (which is operated at ground potential). Actually, however, there are three ways to get negative voltages: (1) by use of coupling capacitors, (2) by accepting input signals from another *electronic box* that happens to have a negative power supply and therefore can provide a negative swing on its output lines, or (3) by accepting an input from an external speed sensor, such as a small alternator, that produces sine-wave voltages. For any of these cases, the inputs may swing negative and turn ON the N^- epi to P-substrate diode, as is shown in Figure 5-17. Notice that *any* N^- epi tub on the IC chip can collect these injected electrons. This can cause strange responses within the IC circuitry. Usually there is no latching in bipolar circuits, and normal circuit operation resumes after the injecting input is raised in voltage and stops acting like an emitter.

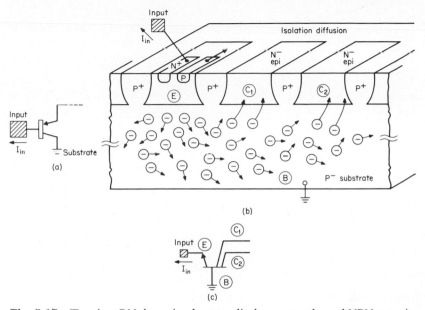

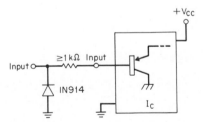

Fig. 5-17. Turning ON the epi-substrate diode causes a lateral NPN parasite: (*a*) the input circuit; (*b*) the lateral NPN parasitic transistor; (*c*) equivalent schematic.

This epi injection can be prevented by making use of external diode clamps, as shown in Figure 5-18. The low-forward-voltage drop of a Schottky diode makes it ideal for this application. An external resistor

Fig. 5-18. Protecting with a diode clamp.

permits use of an ordinary silicon diode and provides a large degree of protection. Resistor values as large as possible should be used if inputs swing negative. Actually, any amount of resistance helps, and large values should be used in harsh electrical environments.

CHAPTER VI

Junction Field-Effect Transistors

The first commercially available field-effect transistors (FETs) were junction FETs, or JFETs. As the name implies, FETs rely on an electric field to control current flow. In this regard, they are more similar to vacuum tubes than they are to bipolar transistors. They are still called transistors, but to be more specific, they should be called "unipolar" transistors. This name results from the fact that only one polarity of carriers, majority carriers, are involved in their operation.

Two types of JFETs are possible: N-channel and P-channel. Since the mobility of electrons in silicon is approximately 3 times higher than that of holes, the N-channel devices are also 3 times faster and have 3 times the g_m.

The JFET has had a dramatic effect on the modern monolithic op amps, and a relatively new JFET structure, the metal-semiconductor FET (MESFET), is a very high frequency device that can provide gain in microwave applications.

As will be seen, FETs can most easily be thought of as a voltage-controlled resistance device in which the input (gate-source) voltage controls the output (drain-source) resistance. There are no problems of recovery time or time delays to establish or change minority charges, as in bipolar devices. In fact, the circuit response time is not usually limited by fundamental processes *within* the FETs, but rather is limited by external-circuit parasitic capacitances and Miller input capacitance (which we will consider in the next chapter). This fact has provided some new FETs with very high frequency capabilities.

Unlike bipolar transistors, FETs use a control voltage (not a base or input current), and when operated at low values of drain-source voltage, they are said to be in their *linear range* because a linear resistance appears between source and drain (the output terminals). FETs are in current *saturation,* where their output impedance is high, when operated with high drain-source voltages, unlike bipolar transistors, which are said to be operating in their *linear range* for the same output-voltage conditions. This nomenclature can cause some initial confusion with

FET devices. The contrast with a bipolar transistor is shown in Figure 6-1. Also notice that the base-current parameter is changed to gate-source voltage on the FET curves.

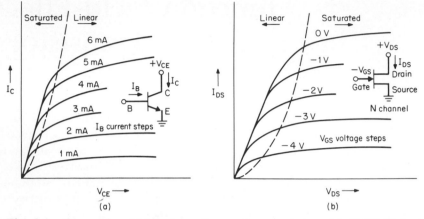

Fig. 6-1. Comparing (*a*) bipolar collector characteristics with (*b*) JFET drain characteristics.

The JFET was proposed by Shockley in 1952 and was first demonstrated by Dacey and Ross. SCLs are fundamental to the operation of the JFET, because a JFET is basically an SCL-controlled resistor. The depletion region of a reverse-biased PN junction is used to modulate and "pinch off" the cross-sectional area by varying the thickness or depth of the channel in which the drain current flows. The channel is of the same doping type as the source and drain regions and is created during the fabrication process. Such a channel is shown in cross-sectional view in Figure 6-2.

Because the channel exists when $V_{GS} = 0$ V, all JFETs are called *depletion-mode FETs:* the application of a gate-source voltage will typically *deplete* the channel and thereby reduce the drain-current flow. Notice that the N$^-$ channel has both a P-type top gate and a P-type bottom gate (the P$^-$ substrate) which we have tied together by allowing the top-gate diffusion to extend into the P$^-$ substrate area. This reduces the channel thickness and will therefore reduce the pinch off voltage, as we will soon see.

The circuits used for the N-channel JFET are similar to those of the NPN bipolar transistor, so we have tied the source to ground and operate the output lead, the *drain*, at a positive-power-supply voltage. The major differences are that the control input is the gate-source volt-

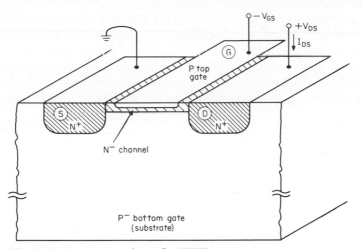

Fig. 6-2.   A cross section of a JFET.

age, which is now negative (at most, equal to zero), and there is essentially no input current.

6.1 THE CURRENT I_{DSS}

One of the characteristic parameters of a JFET is I_{DSS}, the drain-source current that flows when the gate-source voltage is zero. This is the maximum current that will flow from drain to source and is an indication of the size and power capabilities of the device (JFETs designed as amplifying devices for small signals have I_{DSS} values of a few milliamperes). I_{DSS} limits or saturates (in the FET usage of the word) at a relatively low value of V_{DS}.

If we ground both the source and gate leads and consider the sequence of events that happen as V_{DS} is increased in value, we can see the mechanism of operation of the JFET. For small values of V_{DS}, the channel thickness is at maximum because of the small spread of the SCL at low voltages. The channel therefore appears like a resistor, and I_{DS} increases with increases in V_{DS}.

6.2 PINCHING OFF THE CHANNEL

A new pinchoff phenomenon eventually results when the reverse-biasing voltage of the drain-gate junction is finally made large enough. Since there are two increasing SCLs within the channel at the drain end, these can sweep completely across the vertical dimension of the channel (the thickness) and eventually touch each other. This sounds like the end of the FET, and you might expect that with the channel hopelessly cut off, no drain current would flow. The details of this pinchoff condition are diagramed in an enlarged view of the drain end in Figure 6-3. The

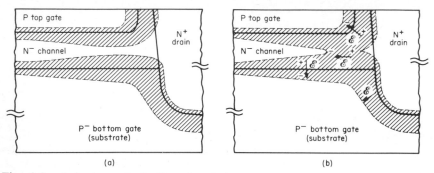

Fig. 6-3. A look at pinchoff at the drain end: (*a*) just prior to pinchoff; (*b*) pinched off.

conditions shown in Figure 6-3*a* are just prior to pinchoff, and Figure 6-3*b* shows the pinched-off state. In both diagrams, the SCL regions are cross-hatched for emphasis.

Our only other encounters with SCL regions had them going across a PN junction. In Figure 6-3*b*, when the channel is pinched-off, there is now an SCL extending from the pinched-off end over to a small region of the channel that is adjacent to the N+ drain region. The direction of the resulting electric field $\mathscr{E}$ is shown within the SCL regions.

Now an SCL is extending from an N-type region to another N-type region. This is similar to reach-through, or punch-through, voltage breakdown in bipolar transistors. Any conducting electrons available within a few diffusion lengths of the channel end are therefore swept across the SCL. The resistance of the remaining channel on the source side of the SCL is the only limit on current flow. This capturing of charge carriers by the SCL is the same as the collector mechanism that takes place in a bipolar transistor, except that the "stolen" electrons in the channel are now majority carriers and there is no PN junction involved here.

The drain current at this pinchoff point is therefore essentially the maximum value that can be obtained and is I_{DSS}. Further increases in V_{DS} have only a slight effect on I_{DSS} because there is some small movement of the SCL down the channel toward the source end; this movement is analogous to Early effect in a bipolar transistor, and the resulting channel-length shortening does cause a slight increase in I_{DSS}.

6.3 V_{GS} CONTROL AND THE PINCHOFF VOLTAGE

In all the preceding discussion, the gate was simply shorted to the source. If we now apply voltages to the gate, we can obtain a useful device. The PN junction between the gate (P) and the channel (N) restricts the positive gate voltages to less than a few tenths of a volt to prevent large forward currents through the diode. In practice, a $V_{DS} = 0$ V is a maximum gate-voltage drive condition.

A reverse-biasing voltage ($-V_{GS}$) on the gate will directly add to the SCL formation in the channel. Therefore, every volt of this reverse bias will reduce the value of V_{DS} at which pinchoff occurs in a $1 : 1$ relationship.

A second parameter of JFETs is the pinchoff voltage V_P. This is the gate-source voltage (negative in this N-channel example) that must be applied to reduce I_{DS} to a near-zero reference value when V_{DS} is small (typical values of V_P are 1 to 2 V). The two parameters I_{DSS} and V_P, therefore, give a good description of the JFET device. The magnitude of V_P gives an idea of the V_{DS} needed to obtain I_{DSS}.

The maximum value of transconductance g_{mo} of a JFET occurs for a V_{GS} of 0 V (g_m linearly decreases to 0 as V_{GS} approaches V_P) and is given by the ratio of $2I_{DSS}$ to V_P, or

$$g_{mo} = \frac{2I_{DSS}}{V_P}$$

For an I_{DSS} of 1 mA and a V_P of 2 V, g_{mo} of a JFET is only 1 mA/V. This is small when compared with the 38 mA/V g_m of a bipolar transistor when also operated at 1 mA of bias current.

6.4 THE BI-FETS*

A major revolution in monolithic op amp design was made possible in 1972 when Ronald W. Russell and James L. Dunkley successfully combined P-channel JFET transistors with the standard linear bipolar process (Figure 6-4). The use of an ion implanter to very precisely put both a channel and a top gate in place has allowed a good match between

---

* Bi-FET is a trademark of the National Semiconductor Corporation.

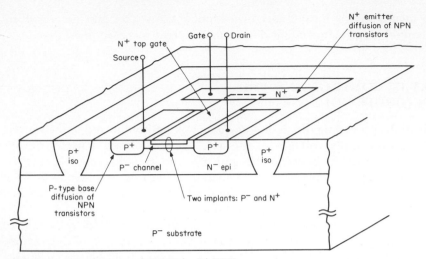

Fig. 6-4. The P-channel JFET in Bi-FET.

devices. The result has been lower input-offset voltages in these Bi-FET op amps than was possible with discrete JFETs added to the front end of standard monolithic op amp chips. Further, the higher-frequency response of the P-channel JFET, when compared with the lateral or vertical PNP bipolar counterparts available in standard monolithic fabrication, has allowed improvements in the high-frequency performance of Bi-FET op amps. This has resulted even though the frequency performance of these implanted JFETs is not as good as that of discrete JFETs. The new process has reduced the cost of adding JFETs. Consequently high-frequency Bi-FETs have essentially taken over the monolithic op amp market. In addition, Bi-FETs facilitated the design of many linear ICs that need the very high input impedance provided by JFETs.

6.5 METAL-SEMICONDUCTOR FETs (MESFETs) AND THE GASFETs

A transistor that operates at microwave frequencies is obtained when a Schottky diode is used for the gate on an N-channel JFET product. This metal-semiconductor FET, called a *MESFET*, is shown in Figure 6-5. A thin layer of N-type GaAs is used as the channel region. The spread of the depletion region of this Schottky-gate diode into the N-type region of the GaAs modulates the drain-source current flow. The structure can be made very small and no diffusions are needed in

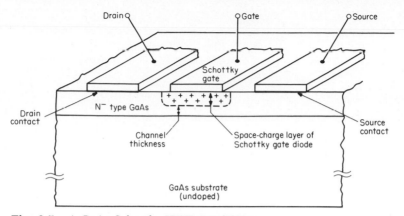

Fig. 6-5. A GaAs Schottky JFET (MESFET).

the fabrication (which is why GaAs can be used). The GaAs substrate is naturally a better insulator than silicon and makes a good supporting structure.

The higher mobility of the electrons in GaAs (approximately 5 times faster than in silicon) has favored this material for high-frequency applications. A competing device is the MOS, or enhancement-mode version, the GASFET. Both these devices are increasing the usefulness of semiconductors in microwave applications: they are able to operate up to 40 GHz with extremely low noise levels, and they also are used for some new, very high-speed logic circuits.

VII

Metal-Oxide Semiconductor Field-Effect Transistors (MOSFETs)

In the past it was rather more difficult to build metal-oxide semiconductor FETs (MOSFETs), but today, large investments in this technology have created more rapid advances than those of the competing bipolar technologies. MOSFETs are the backbone of the microprocessor revolution and the low-cost semiconductor computer memory products. The first commercially feasible MOSFETs were made by Steven R. Hofstein and Frederic P. Heiman in late 1962.

7.1 THE P-CHANNEL MOSFET (PMOS)

As logic densities increased, a process simpler than that of the bipolar ICs was needed to increase yields, reduce power consumption, and provide increased logic complexity at low fabrication costs. P-channel MOS-FETs (PMOS) transistors were the first solution to these problems, and they resulted in *large-scale integration* (LSI) for digital products. Consequently, the PMOS process was responsible for the initiation of many new IC products, such as calculators, complex digital chips, semiconductor memories, and microprocessors.

It took many years of research and development to bring MOFSETs to the market, because their processing had to be "cleaner" than that of bipolar-logic circuits. Sodium contamination occurred in the early PMOS products and caused serious reliability problems: the contaminated FETs could not be turned ON, so circuits would eventually stop functioning. More understanding of the semiconductor surface and better processes have provided stable MOS devices today.

The basic structure of a P-channel MOSFET (Figure 7-1) is obtained by starting with a lightly doped N^- substrate and then using a single P^+ diffusion to locate isolated regions that will serve as either sources or drains for individual PMOS transistors. No expitaxial layer is needed,

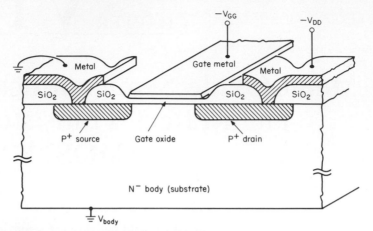

Fig. 7-1. The P-channel MOSFET.

there are no buried layers, and the PMOS transistors are naturally isolated (they can share a common substrate, as we will soon see), so a special isolation diffusion is not needed. This is certainly a simpler process than that used for bipolar circuits.

We will describe basic MOSFET action using the P-channel device as an example. This will allow tracing the development of MOS technologies as they historically occurred. The basic operation of N-channel devices is the same and simply requires interchanging all the dopant types and applied-voltage polarities.

A voltage applied to a piece of surface metal, called the *gate,* which is deposited over a relatively thin (1200 Å) layer of SiO_2 (the *gate oxide*), will affect the surface of the N^- silicon. There are three surface regions of interest in MOSFETs, and these are shown in Figure 7-2 as A, B, and C. The channel region (B) will be described first.

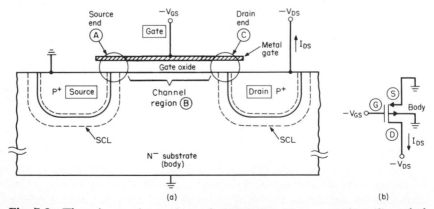

Fig. 7-2. Three interesting regions of a MOSFET: (*a*) cross section; (*b*) symbol.

Inverting the Channel Region

Without a gate voltage, the source and drain regions are isolated from each other (there is no *channel region* of the same doping type connecting them), and therefore no I_{DS} will flow. This provides a desired OFF state for digital circuits and is called an *enhancement mode* in FET operation. The application of a gate voltage will invert the channel region to P-type, which will *enhance* (increase) the drain-source current flow.

There is no direct contact of the gate metal to the surface of the semiconductor. The gate metal is brought close to the surface, but the thin gate oxide forms an insulating layer under the metal. It has been shown that this rather remotely located control element of the MOSFET (like the remote grid of a vacuum tube) reduces the transconductance as compared with the more intimate location of the base region in a bipolar transistor (or the gate region in a JFET with a low pinchoff voltage). For this simple physical reason, the bipolar transistor has the largest ratio of transconductance per milliampere of biasing current. Let us look more closely at the channel region.

MOS Intuition

We can better intuitively understand the action of the gate in a MOSFET by again making use of the concept of a ½ diode, as used in the discussion of the Schottky diode. We will replace the P-type side of the diode with a combination of a metal gate and a thin oxide layer to create an N-MO diode. Because there is an insulating layer (the gate oxide) between the two sides of the diode, a new effect can be obtained under reverse bias: we can actually have mobile carriers *inside* an SCL.

To see how this happens, let us look at the channel region as if it were simply a reverse-biased PN diode, where the P region is replaced by the metal-oxide (MO) layer, as shown in Figure 7-3. As expected, a negative gate voltage will cause an SCL to form at the surface of the N^- substrate. This uncovers fixed positive charge in the lattice, and the negative charge that balances it is now located on the gate. So far, the only new thing is the way we have replaced the P-type side of our diode with the gate metal sitting on the oxide layer.

A second new thing now happens: mobile holes that diffuse into this SCL or are generated within the SCL are *no longer swept out.* These carriers, which would have caused leakage-current flow in a normal PN reverse-biased diode, cannot cross the oxide layer, so they are simply *left sitting at the silicon-oxide interface.* When enough of these holes accumulate (limited by the relatively small values of leakage current) to just exceed the doping concentration at the surface of the N^- substrate,

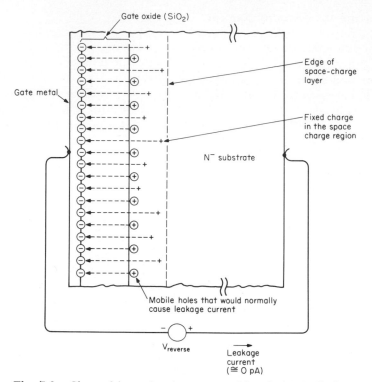

Fig. 7-3. Channel inversion is a reverse-biased N-MO diode.

we have what has been called *weak inversion.* Operation of MOSFETs in this mode is the subject of much current research. This is also called *subthreshold operation* and has been neglected by conventional-logic applications of MOSFETs.

When, with an increase in the gate reverse-bias voltage, the hole density further increases to where the excess is equal to the N-substrate doping density, we have what is called *strong inversion,* the typical minimum operating conditions of a MOSFET. Further increases in reverse-bias voltage beyond the strong-inversion condition bring about additional induced mobile charges and the SCL essentially stops spreading.

So the channel of a MOSFET is seen to be formed from *frustrated leakage current.* If this *current cannot flow up* (through the oxide), perhaps it can be made to *flow sideways* (parallel to the surface oxide): *this is the basis of the MOSFET.* To achieve this requires an additional electric field (which is provided by V_{DS}) at right angles (90 degrees) to the electric field established by the gate voltage. *FETs are therefore crossed-field devices.* When very large gate voltages are applied, the mobility of the charge carriers along the silicon-oxide interface is reduced because of the large density of carriers that large values of gate voltage create. The current in the MOSFET therefore does not increase as rapidly (so g_m falls) at large values of gate voltage. This causes the large-signal current-limiting seen in MOSFETs.

In a MOSFET, the mobile carriers that create the channel (*the channel charge*) are supplied mainly by the source region of the transistor. Without this supply of carriers, it would take a relatively long time for a channel to be formed from just leakage currents. After a channel has formed, if the gate voltage is then rapidly switched back to 0 V, a small number of the holes (the channel charge) will recombine within the substrate region. This gives rise to a *charge-pumping action* when MOSFETs are switched, and this is in addition to the capacitively coupled control-signal problems associated with most electronic switching devices. For example, channel charge is obtained from the source when the transistor is turned ON and a small part of this charge is returned to the substrate (also called the *body*) when the transistor is turned OFF. If a capacitor is tied to the shorted source and drain regions of a MOSFET and the gate voltage is continuously switched from ON to OFF, the capacitor will become charged, as shown in Figure 7-4. The voltage on the capaci-

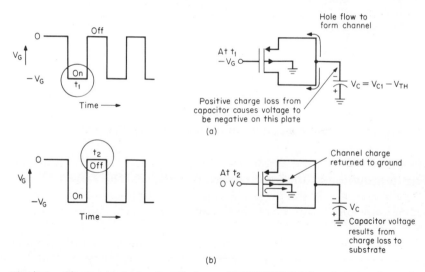

Fig. 7-4. Charge-pumping in a P-channel MOSFET: (*a*) forming the channel at t_1; (*b*) removing the channel at t_2.

tor will reverse for an N-channel MOSFET using a $+V_G$ gate drive. This charge-pumping action causes a small charge loss from capacitors when MOSFETs are used as switches. This can be a design consideration in critical MOS analog circuits.

Conditions at the Source End

When we extend this *½ diode model* to the *source end* (region A in Figure 7-2), we notice that a different diode-biasing condition exists, as shown in Figure 7-5. The typical operating potentials on the MOSFET cause

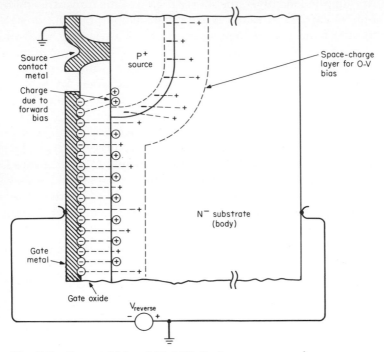

Fig. 7-5. Forward-biased P⁺-MO diode at source end.

this P-MO diode to be *forward-biased* because the P⁺ source diffusion is tied to ground and the gate voltage is negative. The forward current is again *frustrated* by the presence of the insulating gate-oxide layer and again simply piles up at the oxide-silicon interface. This is the *accumulation mode* in the MOSFET.

Notice that this forward biasing of our ½ diode has caused mobile charge to accumulate at the surface of the source region under the gate metal. Those electric field lines from the gate which terminate on the mobile charges that are near the channel region are not perpendicular to the surface. This creates a component of force which eventually causes the charges to be injected into the channel region. *This supplies the channel charge in the MOSFET.* As can be seen in Figure 7-5, the source is ready to conduct as soon as the gate region has accumulated enough channel charge to support a channel current.

Conditions at the Drain End

For small values of V_{DS}, as shown in Figure 7-6, the channel extends to the drain end (region C in Figure 7-2). This represents the *linear region* of MOSFET operation because the drain-source current increases as the drain-source voltage is increased: *the channel is simply a linear resistor.* Note that at small values of V_{DS} the channel charge is uniform from the source end to the drain end. The balancing *mobile negative charge on the gate must always be uniformly distributed* across the gate metal because the high conductivity of the metal will not allow a nonuniform charge

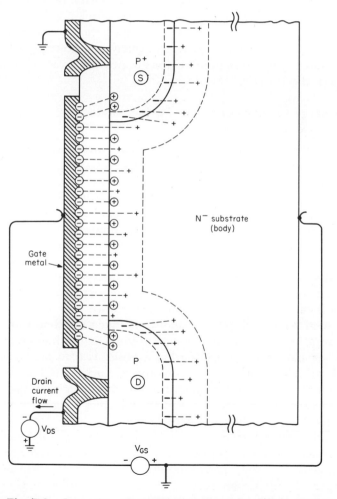

Fig. 7-6. Operation of a PMOS transistor at small V_{DS} voltages.

layer on the gate. Some of this gate charge is used to support the SCL in the N⁻ substrate region. In Figure 7-6, this is shown by the dotted lines drawn from the electrons on the gate to the immobile positive charges in the substrate. To provide for this much charge on the gate, a gate voltage is required and is given by

$$V_{GS} = \frac{Q_{total}}{C_{oxide}}$$

where Q_{total} is the total charge required on the gate, and C_{oxide} is the capacitance between the gate metal and the channel. This relation shows how thin oxide layers, which increase the gate-channel capacitance because of the closer spacing, reduce the required V_{GS} for a given total charge. This is therefore one way to reduce the threshold voltage of MOSFET devices. New processes use many low-doping-concentration implants directly in the channel regions to provide a variety of threshold voltages in the resulting transistors.

Fixed Charge in the Oxide, Fast-Surface States, and Ionic Contamination

Three other sources of charge have to be considered in a MOSFET. One is the fixed positive charge that ends up locked in the oxide layer approximately 200 Å from the silicon-oxide interface. This is called *surface-state charge* Q_{ss} and *is always positive*. Much has been learned about the effects of crystal orientation and processing on the magnitude of Q_{ss}, but why it should exist at all is still somewhat of a mystery: it is believed that Q_{ss} originates from excess ionic silicon that has moved into the oxide during oxidation.

Another charge source is from contamination ions, and the largest problem here comes from the also positively charged mobile sodium ions. Both these additional charges are shown in Figure 7-7, and as can be seen, they both require additional negative charge on the gate to balance them. As a result, the threshold voltage is higher, and PMOS devices are harder to turn ON (the opposite is true for an N-channel FET). Especially at elevated temperatures, early PMOS products that suffered from sodium contamination would therefore eventually not turn ON at all after sufficient time had elapsed for the contaminating sodium ions to migrate into the gate-oxide layers.

A final source of charge is the abrupt ending of the silicon crystal at the surface. A number of silicon atoms at this crystal edge have missing bonds, and as a result, many complex possibilities for the production of charges of either sign exist. This has been called *fast-surface states* because of the larger recombination that exists at the surface. Generally, modern processes have solved the sodium-contamination problem, and

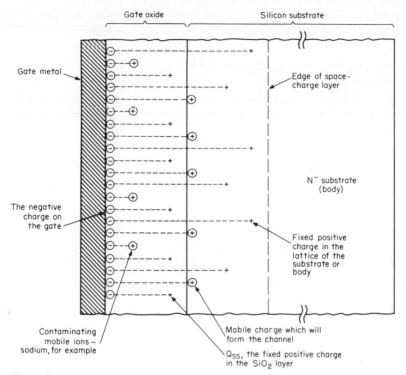

Fig. 7-7. Additional charges involved in MOSFETS.

the charge associated with the fast-surface states is typically added to Q_{SS} in calculating its effect on the magnitude of the threshold voltage. To simplify the explanations, these extra charges were not included in earlier figures and will also not be included in illustrations that follow.

Effects of Gate Material on V_{TH}

A last factor that affects the threshold voltage of a MOSFET is the physics of the gate material. In a manner similar to a Schottky diode, a built-in field will exist at the surface of the semiconductor, and this is due to the presence of the gate material. Since this field affects the surface of the semiconductor, a component of the threshold voltage represents the gate voltage needed to remove the effects of this built-in field (and other effects) and restore equilibrium conditions. This is called the *flat-band voltage*. In silicon-gate processes, which we will consider shortly, the polysilicon gate material is doped (either during the drain and source implant step or separately), and this heavy doping facilitates not only a lower resistance for the gate material but also a desired reduction in the threshold voltage. Modern processes use silicon gates and control threshold voltages with channel implants.

Increasing V_{TH} through Source-Body Reverse-Biasing

All the examples so far have assumed that both the source and the substrate (the body) are operated at ground potential. We will now consider the changes in threshold voltage of a MOS device that result from a body-to-source V_{BS} reverse bias.

In the FET in Figure 7-8, for example, with the source still tied to

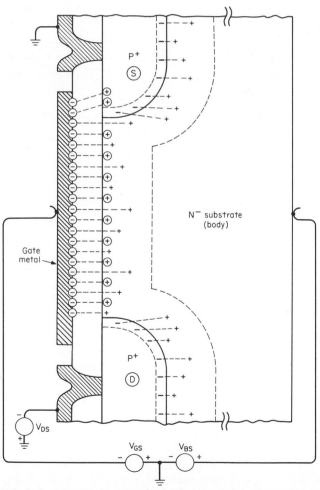

Fig. 7-8. Reverse-biasing the body-source junction reduces channel charge.

ground and the N⁻ body made positive, we have a V_{BS} reverse-bias condition. The SCL across the source-body junction therefore increases in width, as shown in the figure. Notice that the source potential, ground in this example, causes the channel to be at ground at the source end as well. Therefore, the positive reverse bias applied to the body also will cause the SCL down the channel to widen. The fixed negative charge on the gate will now have to accommodate this increase in fixed SCL charge in the body region. This will reduce the mobile channel charge that can be attracted to the oxide-silicon interface by the charge on the gate. Therefore, the V_{TH} will increase if there is a V_{BS} reverse bias.

We could keep the body at ground and, instead, apply a negative voltage to the source lead. Again, this will result in a V_{BS} reverse bias and the SCL from the source to the body will widen. This negative voltage also appears in the channel at the source end and again creates an increase in the width of the SCL into the body. As before, this reduces the channel charge and raises V_{TH}.

Effects of Increases in V_{DS}

When V_{DS} is increased, there is an increase in the width of the SCL that exists across the drain-body junction, as shown in Figure 7-9. Again, the increase in SCL penetration into the body causes a loss in channel charge. This causes the drain-source current to drop and depart from the linear characteristic of the MOSFET that exists at low-source voltages.

The Saturation Region

Finally, for further increases in the magnitude of V_{DS}, a point is reached at which the drain current remains essentially constant. This is called the *saturated region* of operation because *the drain current stops increasing* or reaches a *saturating value.*

The model for this is shown in Figure 7-10. Here we see that the electric field across the drain-body junction is finally intense enough that the gate can no longer maintain mobile channel charge in this region. This causes the channel to end at an SCL, and the drain current is now swept across the SCL in a manner similar to the saturation conditions previously described for the JFET. Saturation in a MOSFET is entered when

$$|V_{DS}| \geq |V_{GS} - V_{TH}|$$

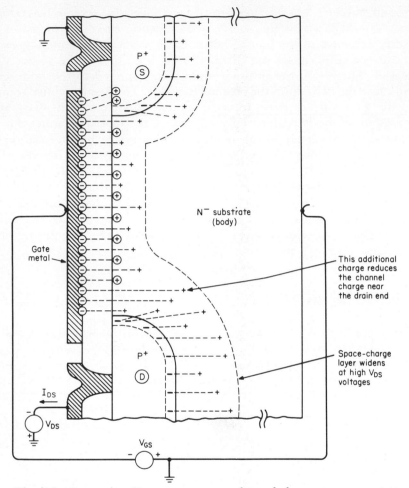

Fig. 7-9. Increasing V_{DS} removes some channel charge.

This indicates that the drain-source voltage is large enough that the SCL at the drain end can successfully compete for the mobile channel charge that exists under the gate at the drain end. Further increases in V_{DS} do slightly increase I_{DS} because of channel-length modulation: the Early effect situation in MOS.

In designing with MOSFETs, one equation is used to predict I_{DS} for nonsaturated operation and a different equation is used for saturated operation. The first step is to approximately determine the biasing voltages of the MOSFET, and this will determine the mode of operation.

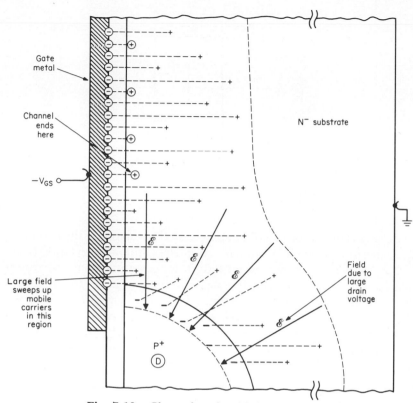

Fig. 7-10. Channel ends with larger V_{DS} voltages.

An Operating MOSFET

To review PMOS transistor action (with the source tied to ground and the drain connected to a negative voltage), current will flow from source to drain under the control of the gate voltage. To obtain conduction, a certain minimum negative gate-source voltage is required to invert the surface to P-type and allow the drain current to flow. This threshold voltage is typically less than 2 V.

Because there are no parasitic PN junctions associated with the gate, large ON voltages can be applied. This is limited by the gate-oxide protection diodes (Figure 7-11) that are used to protect MOSFETs from high values of static-voltage discharge (many kilovolts) onto the input leads. These discharges would otherwise rupture the thin gate oxide and destroy the device. The breakdown field of SiO_2 is 700 V/μm. Therefore, a minimum gate-oxide thickness of 1000 Å (0.1 μm) would break down at 70 V (or less because of oxide imperfections).

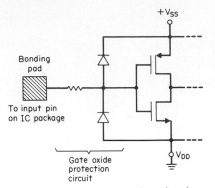

Fig. 7-11. Input-protection circuitry.

The drain current of any FET can be scaled by changing the geometry of the transistor. As expected, from a simple resistor analogy, wide channels that have as short a channel length as can be allowed by drain-source voltage-breakdown considerations will carry large currents. This geometric current-scaling term is stated as W/L (or Z/L) and is the ratio of the effective width of the channel to the effective length of the channel, where the effects of the lateral diffusions of the channel-defining regions are accounted for as shown in Figure 7-12.

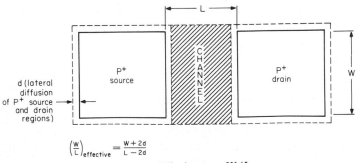

$$\left(\frac{W}{L}\right)_{effective} = \frac{W + 2d}{L - 2d}$$

Fig. 7-12. Effects of lateral diffusion on W/L.

Finally, there is an MOS *gain factor,* sometimes called β', that is a processing-related current-scaling term. The result of all these complications is that many equations must be solved to provide the current drain of a particular MOSFET geometry that has been fabricated with a particular process and is operated at a given set of biasing voltages.

Most processes add an additional step to raise the doping level of the substrate surface in the areas (called *field regions*) where parasitic

PMOS devices are not wanted. This *field doping* therefore prevents the occurrence of unwanted FETs.

The existence of a threshold voltage made MOSFETs a natural choice for digital applications, because the threshold voltage easily allows ON, OFF operation of the transistors. The circuit uses for a P-channel MOSFET are similar to those of the PNP bipolar transistor and therefore do not allow easy interface with standard T²L logic voltage levels. In fact, PMOS products have used negative power supplies (V_{DD} = −12 V_{DC}), and the early products also required a large V_{GG} supply of approximately −18 V_{DC}. Special, logic-voltage-level translator circuitry (T²L to PMOS and PMOS to T²L) was needed at the inputs and outputs of these early PMOS LSI circuits.

A PMOS Inverter

An inverter circuit in standard PMOS is shown in Figure 7-13. The gain device Q_1 accepts an input voltage V_{in} and provides an inverted output voltage V_{out} at its drain. The standard enhancement-mode (non-implanted) load device Q_2 replaces a load resistor for the gain stage.

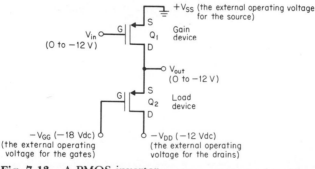

Fig. 7-13. A PMOS inverter.

As a result of the more negative biasing voltage V_{GG} of the gate, this load device will allow an output-voltage swing from ground (V_{SS}) to the full negative-power-supply voltage level V_{DD}. If the extra V_{GG} supply were not available and the gate of the load device were simply tied to the V_{DD} supply, the output voltage would then not swing clear to V_{DD} but would be less because of the threshold-voltage drop of Q_2.

Depletion Loads

An additional ion implantation can be used to selectively add a very low concentration P-type dopant directly into the channel regions of some of the transistors. These transistors will be used as *depletion-load devices* in logic circuits. The channel implant removes the requirement for a threshold voltage, since these devices will have a channel (because of the implant dopant) and will conduct a current with a gate-source voltage of 0 V. Depletion-load devices eliminated the requirement for the additional V_{GG} power supply and also improved switching speeds. Use of a depletion-load transistor is shown in the circuit in Figure 7-14. Now the output voltage will swing clear to V_{DD} and the V_{GG} power

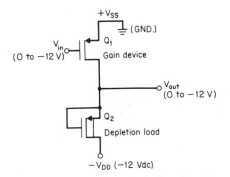

Fig. 7-14. Using a depletion load in a PMOS inverter.

supply can be eliminated. As expected, the depletion-load transistor also has been useful in N-channel products. Load devices which conduct dc current when the gain stage is ON, add to the power consumption of the logic chip.

Improvements with Silicon Gates

The basic PMOS process has varied greatly over the years. One of the more interesting variations was elimination of the metal gate and use of a polysilicon layer deposited on the wafer and then heavily doped during the source and drain diffusion. This polysilicon layer (aided by a nitride cover) kept the dopant out of the silicon and therefore separated the source and drain regions and defined the gate regions. This *silicon-gate technology* (Figure 7-15) allowed not only a reduction in the threshold voltage (as a result of the use of an impurity-doped polysilicon layer), but also provided a *self-aligning gate* that raised the frequency performance of the circuits by approximately a factor of 5 : 1.

In silicon-gate processing, the polysilicon gate regions are formed first and then are used to provide a dopant mask for the location of the source and drain regions. This eliminates the problems of the metal-

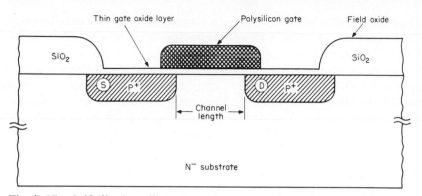

Fig. 7-15. Self-aligning silicon gate determines channel length. P⁺ source and drain dopant penetrates the thin gate-oxide layer but is masked by the polysilicon gate material.

gate processes that require the gate metal to overlap both the source and drain regions to ensure that the channel will connect to both, even when there is a misalignment of the masks used in wafer fabrication.

Unfortunately, this gate-metal overlap increases the input capacitance because of *Miller effect,* and this limits the speed of the metal-gate products. Miller effect occurs when parasitic capacitance from the input to output in an inverting-gain stage (the logic inverter) appears very large in value at the input (Figure 7-16). This happens because a small voltage change at the input causes a much larger voltage change at the output (owing to the voltage gain of the stage). Miller effect

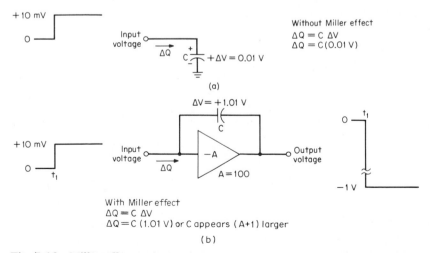

Fig. 7-16. Miller effect makes capacitors "appear" larger: (*a*) supplying charge to a capacitor; (*b*) Miller effect.

increases the *total voltage change across the capacitor.* As a result, the signal source (at the input) must supply an increased current to this capacitor, because the capacitor "appears" larger by this gain factor. Miller effect is a problem in most high-speed circuits. The self-aligning feature of the silicon-gate process removed the large overlap capacitance of the gate in the metal-gate process. This reduced Miller input capacitance and increased speed. Further reductions in overlap capacitance have been achieved by special processing (in the modern NMOS products) that actually creates a *negative overlap* (the source and drain regions *are not quite covered by the silicon gate*). Connections to the channel region are provided by the fringing action of the electric fields of the MOS device. As expected, this provides higher-speed circuits.

Today the PMOS process is not being used for any new general-purpose logic; it has been replaced by the newer N-channel processes.

A new load possibility was made available when a process was developed that produced both P-channel and N-channel devices at the same time. This complementary MOS idea will be described next.

7.2 THE COMPLEMENTARY MOSFET (CMOS)

The complementary MOSFET process (CMOS) has both N-channel and P-channel MOSFETs available on the same chip. The use of CMOS has permitted the design of some exceedingly low-power-drain logic circuitry. This results from the use of a stage where no dc current is allowed to flow which has become the basic inverter of and the central idea for other CMOS logic gates. The inverter stage (Figure 7-17) has only the upper side (P-channel) or the lower side (N-channel) ON at any given time, so *there is no dc current flow.* For V_{in} at 5 V_{DC}, the

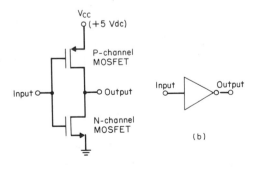

(a)

Fig. 7-17. A CMOS inverter (no dc power drain): (*a*) the circuit; (*b*) the logic symbol.

P-channel is OFF and the N-channel is ON. This interchanges for V_{in} = 0 V. Actually, both transistors are ON during a small part of the transitions of the input-voltage wave form, and in addition, stray capacitances are charged from the power-supply voltage source and then discharged to ground once every clock cycle of operation. These factors cause an operating-frequency(clock frequency)-dependent power consumption in CMOS logic circuits.

The low power drain of CMOS logic circuits is needed in watches and many other special applications, such as pocket calculators and other battery-powered systems. Moreover, low-standby-drain, static, random-access memory (RAM) is a natural choice for the power-saving CMOS technology. CMOS circuits can be operated from power-supply voltages as low as a few volts (for watch applications) to as high as 15 V, but operating CMOS logic from a single 5-V_{DC} supply eases the problems of interfacing with standard T²L-logic voltage levels.

The CMOS process is complicated because a P^- *well* has to be implanted into the N^- substrate to form the regions where the N-channel transistors can be made. In addition, an N^+ diffusion has to be added to form the source and drain regions for the added N-channel transistors. A simplified profile of a logic inverter is shown in Figure 7-18. The

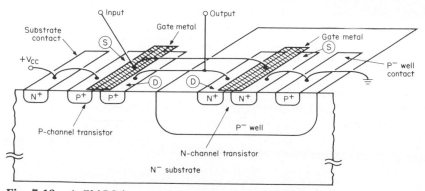

Fig. 7-18.   A CMOS inverter in simplified cross section.

figure does not show the extra locations of P^+ and N^+ diffusions used to prevent parasitic MOS action by raising the surface concentrations of both the field regions. The field regions of the P^- well are thus converted to P^+ and the field regions of the N^- substrate are converted to N^+ by making additional use of the P^+ and N^+ diffusions. The newer CMOS processes use field implants of both types to prevent parasitic MOS devices from appearing in unexpected places within the layout. Use of these implants increases the logic density of the new processes and also eliminates the 7-V maximum-supply-voltage limitation that existed because the P⁺ and N⁺ regions were allowed to touch in the older high-density *butted-guard-band* layouts.

There Is a Vertical NPN Bipolar Transistor in CMOS

If we call the P⁻ well a base region, the N⁺ inside this well an emitter, and the N⁻ substrate a collector, we have a *parasitic NPN bipolar transistor* that has its *collector tied to the power-supply line.* This makes a useful emitter follower and can have a β of 3000. Such emitter followers have been used to provide hundreds of milliamps of output current for LED displays and other power applications.

The Semiconductor-Controlled Rectifier (SCR) in CMOS

A lateral PNP bipolar transistor is also possible where a P⁺ source or drain region of the P-channel MOSFETs can act as an emitter and a P⁻ well can serve as a collector. Therefore, both NPN and PNP devices exist, and these two bipolar transistors can make up a "hook connection," or semiconductor-controlled rectifier (SCR), that can latch and draw large destructive currents. Such an SCR is shown in Figure 7-19. To

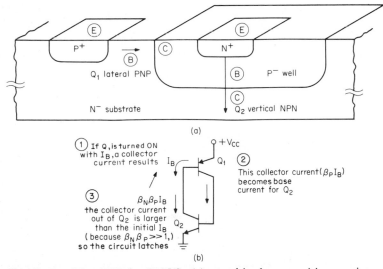

Fig. 7-19. The SCR in CMOS: (*a*) two bipolar parasitic transistors: (*b*) the SCR or "hook" connection. Assume Q_1 is turned ON and a base and collector current results as shown. Loop gain of $\beta_n\beta_P$ causes latch, once started. (β_n is the β of the NPN transistor and β_P is the β of the PNP transistor.)

activate this structure requires that the voltage applied to a lead from the CMOS chip must be taken either below ground or above the positive-power-supply voltage level. This caused problems in early CMOS products, but techniques are available today to modify the layouts and essentially eliminate this parasitic bipolar problem.

The Transmission Gate

Another benefit comes with CMOS: the *transmission gate* (Figure 7-20). This switch can operate with input voltages that range from 0 V_{DC} to

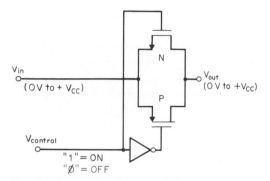

Fig. 7-20. A CMOS transmission gate.

V_{CC}; it therefore makes a good linear switching element. It is also useful in logic design and causes CMOS logic to be different from other forms of logic. For example, a CMOS flowthrough latch is shown in Figure 7-21. This is a low-component-count design and makes full use of trans-

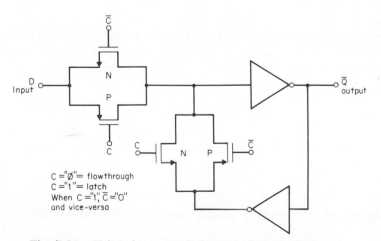

Fig. 7-21. Using the transmission gate in logic design.

mission gates. When the clock C is low (logic 0), the data present at the D input will flow through and the logic inverter will provide the $\bar{Q}$ output. When the clock line goes high, the last state of $\bar{Q}$ is latched and the D input is disconnected.

Using CMOS for Linear Functions

The benefits of CMOS for both analog-to-digital (A/D) and digital-to-analog converter (DAC) products have resulted in their use in many new designs. The processing for these circuits must solve linear high-accuracy problems as well as provide high densities for the required digital-logic circuits. The CMOS transmission gate can switch analog voltages that range from 0 V to the V_{CC} supply-voltage level. These electronically controlled switches are useful at the analog input of an A/D as well as in the resistive ladders of DACs.

The increased variety of doped-silicon regions in the CMOS processes also provides greater opportunities for unusual devices which are useful in linear-designs. Parasitic bipolar NPN transistors are available, 7-V "zener" diodes can be realized, and the existence of both N-channel and P-channel devices is similar to the NPNs and PNPs of bipolar linear IC chips.

Although existing linear circuits can be converted to CMOS by replacing the NPN bipolar transistors with N-channel MOSFETs and the PNPs with P-channels, this does not provide optimum use of the new technology. More benefits result if the basic design approach is changed from that of standard linear circuits, which are based on always having an input signal (or *continuous systems*), to that of *sampled-data systems*, which only periodically sample the input signals. Periodic sampling is the way people operate when they drive cars or do most things. For example, a visual sample of the position of the car is taken by the driver with respect to where the car should be (based on white lines and so forth). A correction to the system is made, via the steering wheel, and the driver is then free to do other tasks, such as observing the scenery.

Analog-to-digital converters (A/Ds) based on this sampled-data concept, are natural uses for CMOS. In general, sampling requires electrically actuated switches. MOS technologies offer excellent switches, and large numbers of them can be used economically in an A/D design.

To gain a better feeling for sampled-data circuits, we will consider a basic *voltage comparator*, as shown in Figure 7-22. The switches SW_1 and SW_3 are CMOS transmission gates and therefore can handle 0- to 5-V input voltages. SW_2 is simply an N-channel transistor. Gain for this comparator is provided by a logic inverter. The transfer function of

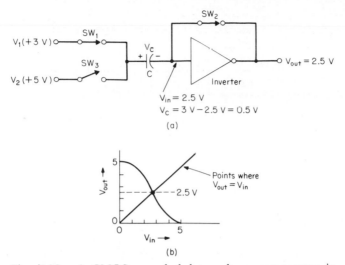

Fig. 7-22. A CMOS sampled-data voltage comparator in the sample mode: (a) comparator; (b) inverter transfer function.

the inverter is shown in Figure 7-22b. For V_{in} close to 0 V, $V_{out} = 5$ V and there is no gain at this operating point. Similarly, for V_{in} close to 5 V, $V_{out} = 0$ V, where there is also no gain. Note that when $V_{in} = 2.5$ V, the circuit has the largest gain, because small changes in V_{in} around this 2.5-V operating point give large changes in V_{out}. Connecting the output to the input of this inverter forces the inverter to have $V_{in} = V_{out}$. As shown in Figure 7-22b, when this occurs, $V_{in} = V_{out} = 2.5$ V.

While SW_2 is closed, SW_1 is *sampling* the comparator input voltage V_1 (3 V). The input capacitor C therefore charges to the difference between 2.5 V and V_1 (3 V) because it is connected between them. In other words, the voltage V_C on the capacitor is 0.5 V. The capacitor will now serve as a memory for the value of V_1 (as well as the 2.5-V inverter reference voltage). After the capacitor has fully charged (to 0.5 V), internal timing signals cause SW_1 and SW_2 to open. Notice that the inverter is now at its high-gain point and will respond to any deviation in the input voltage V_{in} that makes it different from the 2.5-V reference value.

Switch SW_3 is now closed. If V_2 were equal to V_1, for example (it is not in this example), there would be no change in the voltage at the input to the inverter. As shown in Figure 7-23, SW_3 is connecting 5 V (the V_2 input) to the capacitor. The value of V_{in} (to the inverter) is now 5 V (V_2) minus the voltage stored on C ($V_C = 0.5$ V), or $V_{in} = 4.5$ V. Such a large input voltage greatly exceeds the 2.5-V reference

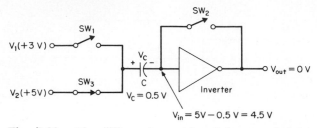

Fig. 7-23. The differencing cycle of the sampled-data comparator.

and will cause the inverter to provide a V_{out} of 0 V, which signals that V_2 is greater than V_1. Note that if V_2 were, instead, equal to 0 V, V_{out} would be 5 V, signaling that V_1 is greater than V_2.

More capacitively coupled inverter stages can be cascaded (each with a similarly timed shorting switch, such as SW_2) to obtain a larger overall gain for a more useful comparator. The thing to observe here is the basic idea of sampled-data circuits and how switches can be applied to useful circuit functions.

High-Performance CMOS (HCMOS)

The developments in CMOS processing are following those in N-channel processing. Today there are fewer differences between these processes, and this reduces the added wafer-processing costs usually associated with CMOS.

Logic densities equal to those of N-channel processes can be achieved by designing all the logic with only N-channel devices and then placing these in one large P^- well. This has been called the *ubiquitous-well concept* and requires depletion-mode N-channel transistors, a new requirement for these CMOS circuits, or other load devices because P-channel transistors are not available in the P^- well.

The low power drain of CMOS makes it viable for reducing dissipation in VLSI products. Power dissipation can be reduced by at least a factor of 10 when compared with NMOS. Newer CMOS processing is catching up with that of NMOS: channel lengths of 0.5 μm can be expected in the not too distant future.

To be compatible with erasable programmable read-only memory (EPROM), a new N^- well CMOS (also called NMOS/CMOS) process has been used. This has a P-substrate where the N-channel transistors are located, and an N^- well is used for the P-channel devices. Although all the doped regions are reversed from those of standard CMOS processing, there are no external changes needed in the application circuits.

To save die area, a form of merged CMOS has been devised that makes use of buried-load devices. This has been called *buried-load logic* (BL²), and the new load device is the parasitic N-channel JFET that can be formed between adjacent P⁻ wells, as shown in Figure 7-24*a*.

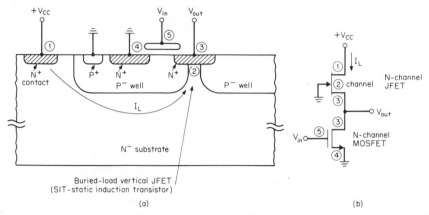

Fig. 7-24. Buried-load logic: (*a*) structure; (*b*) circuit.

Note that the logic inverter in Figure 7-24*b* is now essentially only one transistor in area because of the buried JFET load device. BL² is used for high-speed circuits (it consumes dc power), and normal CMOS is still used for the low-power, slower-speed section of a large chip to reduce the power drain and prevent excessive die temperatures.

CMOS Silicon-Gate Double-Poly (P²CMOS)

Silicon-gate CMOS processes are being used to fabricate random-access memory (RAM) and microprocessor products. These designs make use of small geometries and multiple deposited layers of heavily doped polysilicon to provide not only self-aligning gates, but also multilevel conductors for increasing the logic density. Such a P²CMOS process is very similar to the new N-channel processes we will discuss later in this chapter.

Silicon on Sapphire

The crystal structure of sapphire (aluminum oxide, an insulator) closely resembles that of silicon. Because of the similarity, a thin silicon layer can be epitaxially grown on a sapphire substrate, as shown in Figure 7-25. Further, this epi layer can then be sectioned into separate electri-

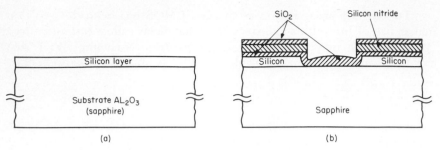

Fig. 7-25. Silicon-on-sapphire isolation technique: (*a*) silicon epi growth; (*b*) masking and complete separation of silicon epi regions by oxidation.

cally isolated regions by selectively *oxidizing completely through* the layer, as shown in Figure 7-25*b*. Junction diodes are no longer needed for electrical isolation of the individual transistors, and the parasitic capacitive coupling from each device to ground is greatly reduced.

Consequently, silicon on sapphire provides *the ultimate in low-stray-capacitance circuitry.* Usually, CMOS processing is combined with this isolation technique to obtain high-speed, low-power-drain circuits, and this is called *CMOS SOS* (silicon on sapphire). (Conventional CMOS is often called *bulk CMOS.*) However, such complex technology is not a high-volume process with major semiconductor manufacturers, but it is used by some of the system and instrument manufacturers because it provides higher speed for those applications which tolerate the extra cost of the sapphire wafer.

7.3 DOUBLE-DIFFUSED METAL-OXIDE-SEMICONDUCTOR (DMOS) TRANSISTORS

Another way to build a MOSFET is to establish the channel length as the difference between the lateral diffusions of a P-type and an N-type region. Both diffusions use the same oxide window (the etched opening) to prevent masking misalignment. Such a MOSFET has been called the *DMOS,* or *double-diffused MOSFET,* and a cross section is shown in Figure 7-26. The short channel length is achieved by differential lateral diffusion without the requirement for expensive masking. A separate mask step is used to locate the N^+ diffusion for the drain. The relatively light P^- substrate (sometimes called a π-region) is easily inverted by the Q_{ss} charge in the oxide layer, and this also allows the use of depletion-load devices in the design of logic inverters.

The rapid progress in masking technology has reduced the relative benefits of the DMOS structure for logic applications, and it is not competitive with the newer high-performance N-channel processes. How-

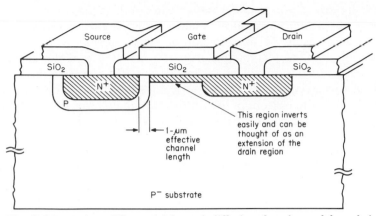

Fig. 7-26. Using differential lateral diffusion for channel length in DMOS.

ever, the DMOS structure is being used in high-power discrete FETS and has allowed many interesting new products.

7.4 VERTICAL METAL-OXIDE-SEMICONDUCTOR (VMOS) FIELD-EFFECT TRANSISTORS

The VMOS transistor (Figure 7-27) is another way to provide a short channel length without expensive masking. Note that the channel for this structure is established by the thickness of the P-layer, which is

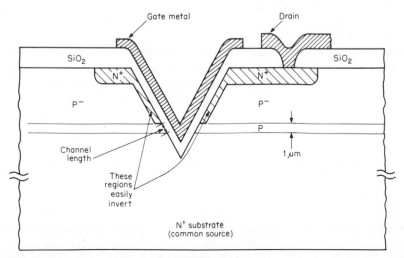

Fig. 7-27. Cross section of a VMOS device.

epitaxially grown and is under better control than the differential diffusions of the DMOS structure. Again, the P⁻ region is easily inverted and can be considered an extension of the drain region.

The groove, or pit, for this structure makes use of the anisotropic etch characteristics of silicon (when in a <100> crystal orientation), where a very accurate 54-degree etch-pit angle from the horizontal is maintained to the bottom of the pit (where the etching stops). A further benefit is that the width of the channel extends on four sides around the pit circumference. Finally, no metallization is needed for the common source and this saves surface area.

Such a VMOS structure also has been used for high-power MOSFET discrete transistors. With FETs there are no safe operating-area limits, and the FET tends to naturally limit current at high temperatures. Power FETs are an interesting newcomer to the marketplace and there is no reason that further developments in power FET technology, *even power IC FET technology,* cannot be made. Again, the N-channel processes are more attractive for ICs, but the VMOS and DMOS technologies have found applications as discrete power devices.

7.5 N-CHANNEL SILICON-GATE MOSFETs

New N-channel silicon-gate processes have overtaken all competitors in memory products and are even challenging the bipolar technologies. NMOS production is dominated by memory and microprocessor products.

Extremely dense logic has been made available through improvements in mask-making (the use of projection masking and direct wafer writing) and dry-etching techniques (plasmas in a vacuum system instead of wet-chemical etches). The new, very large scale integration (VLSI) geometries are so small that channel lengths are approaching the base widths of bipolar transistors (0.5 to 2 μm) and gate oxides are reduced to 400 Å. This is opening the way for the next generation of logic circuits, which will operate on less than 5 V_{DC} (which is *too high a voltage for these short-channel devices*). The 2- to 3-V_{DC} range is therefore being considered as the next standard logic supply voltage, but it is expected to take many years to eliminate the 5-V standard.

The increased mobility of electrons, as compared with holes (approximately 3:1), favors N-channel devices. Unfortunately, the positive trapped charge Q_{ss} that exists in the oxide layer near the silicon surface (which naturally favors the operation of the P-channel enhancement MOS) and the positive charge of most undesired ionic contaminants tend to cause N-channel devices to convert to the depletion mode of operation. This was the initial reason for the substrate-biasing circuits in use with N-channel products: to control and guarantee a threshold voltage. In single 5-V systems, a 0-V logic signal will not turn OFF an N-channel transistor if it has converted to a depletion-mode device; therefore, circuit operation will stop without a control on threshold voltage. Newer N-channel processes have eliminated this requirement for substrate biasing but still take advantage of the increase in punch-through voltage that results when short-channel devices are operated with a substrate bias.

New Uses for Polysilicon Layers and Metal Silicides

In the newer silicon-gate technologies, additional polysilicon layers (up to three) are used as multilevel interconnects. These silicon layers are deposited onto the surface of the silicon wafers by using a chemical-vapor-deposition (CVD) process. The layers are actually polycrystalline films because they are deposited on the amorphous surface oxide of the wafer. Heavy doping of these layers can provide useful conductors (25 $\Omega/\square$, where $\Omega/\square$ is read as "ohms per square," and the number of squares associated with a resistor is simply the geometric ratio of the length to the width of the resistor) that increase the circuit densities. For comparison, the sheet resistance of aluminum interconnect metal that is 1 μm thick is 0.028 $\Omega/\square$. Very lightly doped polysilicon layers also are used to create large-valued resistors, and these are being used in high-density static-RAM designs (which we will also consider in this chapter).

Multiple polysilicon processes *yield better* (that is, they provide more good die on a wafer) than multilayer aluminum processes, but the relatively large resistance and stray capacitance of the distributed RC polysilicon lines causes undesirable time delays in high-speed circuits. The solution has been to modify the polysilicon layer by alloying it with a metal to produce a metal silicide. Silicides are binary compounds of silicon with another element. Metal-silicide films of 3 $\Omega/\square$ or less are replacing the interconnect polysilicon lines in the processes used to fabricate high-speed circuits.

Fabricating NMOS Transistors

The current NMOS process is extremely advanced when compared with the simplicity of early metal-gate PMOS wafer fabrication. Two ion-implant steps (at least) are used; both depletion- and enhancement-mode transistors are provided (with many different threshold-voltage possibilities); one to three polysilicon layers are available; noncontact projection mask aligners are used; and the more precise dry-etching technologies have replaced wet-chemical etching.

The steps involved in fabricating N-channel transistors are shown in Figure 7-28. A starting P⁻ substrate is first oxidized to form the gate

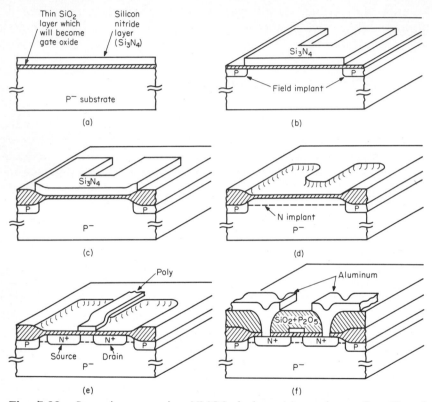

Fig. 7-28. Steps in processing NMOS devices: (a) starting wafer; (b) etch silicon nitride and implant P field; (c) grow SiO₂; (d) remove silicon nitride and implant N depletion channel; (e) deposit poly, etch poly, deposit oxide (not shown), and implant source and drain regions (this is a depletion MOS); (f) deposit thick SiO₂, open contacts, deposit aluminum, and etch aluminum.

oxide on the complete surface of the wafer. Next, a thick layer of silicon nitride (Si₃N₄) is deposited (Figure 7-28a). This nitride layer is an implant barrier and oxidation mask.

The nitride layer is then mostly removed, being left only over those regions which will become sources, drains, or channels of the NMOS transistors or N-type resistors. A P-type ion-implant step then allows penetration of the exposed thin gate-oxide regions and raises the surface concentration of the substrate to prevent unwanted parasitic N-channel MOSFETs (Figure 7-28*b*). This field implant is also entered into the channel regions of enhancement-mode FETs (not shown).

A thick layer of field oxide is then grown (Figure 7-28*c*). Following this, all the nitride is removed from the wafer. An N-type implant layer is then introduced into all the transistor regions (Figure 7-28*d*).

Polysilicon (sometimes simply referred to as *poly*) is deposited over the complete wafer, which is then doped. Most of the polysilicon is then removed, leaving only the gate regions of the transistors and other polysilicon interconnect lines. A relatively thin oxide layer is then deposited prior to the N^+ source-drain implant (Figure 7-28*e*).

Additional oxide is deposited along with phosphorus pentoxide (P_2O_5). The phosphorus acts as a *getter* for any mobile sodium ions that may be present. (Sodium is attracted to the phosphorous atoms and becomes trapped at their locations, thus removing them, or *gettering them,* so they will not affect the transistor operation.)

A cross-section of both depletion- and enhancement-mode transistors is shown in Figure 7-29. The depletion-mode device is achieved

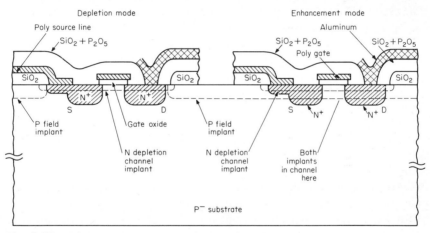

Fig. 7-29.   The NMOS transistor in cross section.

by inverting the channel region with an N-type ion-implanted layer. The polysilicon line shown connected to the source was deposited directly on the exposed surface of the silicon. When this polysilicon layer is doped, the N-type dopant actually penetrates the layer and contacts to and extends the source region, as shown. The polysilicon gate is contacted by a separate metal line, which, for clarity, is not shown. An aluminum line also contacts the drain region and sits on top of the deposited SiO_2-P_2O_5 doped oxide layer.

The enhancement-mode device is similar, except notice that the P-field implant also was placed in the channel region. This prevents the N-channel implant from inverting the surface and thereby provides an enhancement-mode device.

7.6 NMOS IMPROVES MICROPROCESSORS

If we look back in time, we find that calculator chips preceded the microprocessor. In early 1971, the complete logic for a four-function calculator was integrated on a single PMOS chip. Later that year, the first microprocessor appeared—the 4004. This microprocessor also was built with PMOS technology. The later change to NMOS greatly eased the T²L interface problems.

The early NMOS products required three power-supply voltages, 12, 5, and −5 V. Improved processes have since reduced this to a single 5-V supply. The emerging N-channel processes could not only more easily meet the T²L input voltage and load specifications, they also reduced the chip dissipation from about 85 mW per output buffer (with PMOS) to about 5 mW per output buffer. In addition, they increased speed by a factor of 6:1.

The power savings were used to permit the addition of increased capabilities, because the design of a fast central processing unit (CPU) is more limited by power dissipation than by chip size. Therefore, additional functions and a reorganization of the CPU provided a 10:1 speed improvement in computing. This was a larger factor than the 3-times-mobility benefit of N-channel devices.

Microprocessor chips reduced the requirement for custom logic and provided a second high-volume product family for the emerging LSI technology (in addition to memory). The use of read-only memory (ROM) products was increased because of the appearance of the microprocessor. This fixed storage medium is typically used for the program of low-cost digital computers/controllers. (General-purpose computers use RAM storage, with a minimum of ROM, so they can be easily reprogrammed for various uses.)

To better understand N-channel microprocessor products, we will take a look at some of their logic circuits. We will not get into CPU tradeoffs, but we will see how processes have affected the design of logic circuits.

NMOS Logic Circuits

The basic logic form for NMOS is the NOR gate, shown in Figure 7-30. An inverter results if only a single input is provided. A depletion

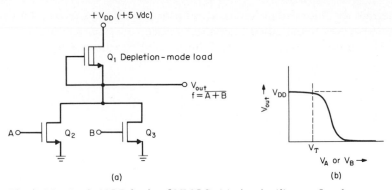

Fig. 7-30. Basic NOR logic of NMOS: (*a*) circuit; (*b*) transfer characteristic.

load is used for the pull-up, and the logic swings are from 0 to 5 V. The transfer function of the NOR gate is also shown in this figure. This simple NOR gate can be used for on-chip logic, but is not usually used to drive load capacitances larger than about 1 pF.

A layout of a single-input gate, the inverter, is shown in Figure 7-31. The contact of the polysilicon layer with the N⁺ region is obtained simply by removing the SiO₂ in the contact area prior to deposition of the polysilicon. When the layer is subsequently doped, the dopant will

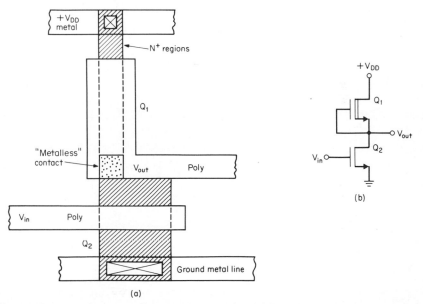

Fig. 7-31. Layout of an NMOS logic inverter: (*a*) plan view; (*b*) circuit schematic.

penetrate through the polysilicon and into the silicon surface. This provides what has been called a *metalless contact*.

The buffer shown in Figure 7-32 is used for driving heavier loads.

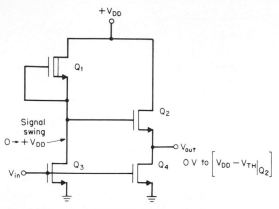

Fig. 7-32. An NMOS inverting buffer.

Notice that the depletion device Q_1 drives the upper output transistor Q_2, which provides a drive voltage that swings clear to the V_{DD} level. However, there is a V_{TH} drop across Q_2 that limits the output-voltage swing to $V_{DD} - V_{TH}$. In general, the output transistors Q_2 and Q_4 are large W/L devices that increase the output-current capability of the buffer. This undesirably raises the input capacitance of this stage.

As expected, NMOS logic makes extensive use of depletion loads. The extra current available to charge ouput capacitances with a depletion load is shown in Figure 7-33. Here we compare a resistor load, a deple-

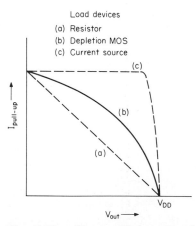

Fig. 7-33. Characteristics of various load devices.

tion load, and a current-source load. The depletion load is an improvement over the resistor load, but it does not provide as much current as a more ideal load device, a true current source. In a logic inverter, as the input transistor turns OFF, the load device must charge up the total capacitance that exists at the output node. With a resistor load, as the output voltage increases, the drop across the resistor decreases and this reduces the available current. Note the improvement with the depletion load.

The drain characteristics of a depletion-mode MOS transistor are shown in Figure 7-34. This shows that approximately 600 μA of current

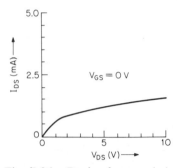

Fig. 7-34. Drain characteristics of depletion-load transistor.

is available with a $V_{DS} = 1$ V. Moreover, a maximum current of 1.5 mA can be obtained at $V_{DS} = 10$ V. In a load device, it is desirable to have large currents at low V_{DS} that do not increase significantly as V_{DS} increases.

Such a depletion-mode characteristic can be compared with the enhancement-mode curves of Figure 7-35. The g_m at any operating point

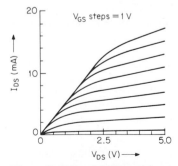

Fig.   7-35.   Enhancement-mode NMOS transistor drain characteristics.

can be determined by noting the change in drain-source current obtained with a small change in gate-source drive voltage. The ratio of these two quantities is the g_m of the transistor. The output impedance of the transistor also can be determined by taking the reciprocal of the slope of a constant V_{GS} curve, or $r_{out} = \Delta V_{DS}/\Delta I_{DS}$, with V_{GS} constant. Finally, the value of V_{DS} for a given I_{DS} and V_{GS} drive can be determined from these characteristics. This is important to ensure meeting a T²L-logic 0-output-voltage specification: 0.4 V at 1.6 mA of output current.

MOS output buffers that can drive one T²L load present size and chip-dissipation problems. Some designs reduce dissipation by driving one low-power T²L load (0.18 mA at 0.3 V) or multiplexing the output buffers to reduce the total number needed. Multiplexing has been used with many of the newer microprocessors and results in reduced chip size, less power dissipation, and fewer package pins; however, it increases the complexity of the user's interface.

7.7 CHARGE-TRANSFER DEVICES (CTDs)

The fact that the inversion charge of an MOS device is slow to accumulate if no source region is available has allowed the design of some unusual NMOS products. These products achieve various useful electronic functions by *passing inversion-charge packets* from device to device, and, they are all generally called *charge-transfer devices* (CTDs).

Much research activity has centered around two basic circuit types of charge-transfer devices: the charge-coupled device (CCD) and the bucket-brigade device (BBD). Many digital applications have been thought of, such as shift registers and memory, but the main uses are in analog-signal processing (filters) and charge-coupled imagers (CCIs).

Bucket-Brigade Devices (BBDs)

If we recall that the basic MOS transistor is symmetrical and the source and drain can be functionally interchanged, we can understand the basic concept of the bucket-brigade digital-shift register shown in Figure 7-36.

To start, imagine that there are no initial charges in the circuit and that all the gate voltages (V_2 through V_5) have just been taken positive and are all equal to or greater than the threshold voltage. We will use two magnitudes of gate voltage, a minimum that is just greater than V_{TH} and a maximum that is greater than this. The channel regions under G_2 to G_5 are all biased for inversion, but the only source of mobile electrons is from the P-substrate (for now, we are neglecting the gate

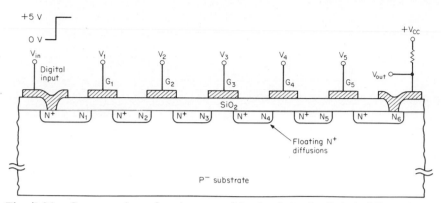

Fig. 7-36. Cross section of an integrated bucket-brigade digital shift register.

voltage V_1, assumed to be 0 V, and the digital input voltage V_{in}). Depletion layers are formed under each gate, but *initially no minority carriers (channel charges) are available.* Electron-hole pairs are thermally generated in the SCLs, and if this is the only supply of mobile electrons, it would take several seconds for the inversion layers to form. However, if V_{in} is 0 V and V_1 is more positive than the voltages at the other gates, electrons can be rapidly drawn from N_1 to form a channel under G_1. This *launches a signal* at the input to the bucket-brigade (and is the same way switched MOS devices get their channel charge). If V_{in} were, instead, 5 V, there would be no channel formed under G_1. In this way, we can launch a charge packet or not, depending on the state of the digital input signal V_{in}.

With V_1 more positive than V_2, any mobile electrons under G_1 (if V_{in} is 0 V) will not be lost to G_2. Now if we also make V_2 a maximum voltage and then return V_1 to a minimum voltage, the charge packet that formed the channel under G_1 will almost entirely transfer to the new, deeper depletion region under G_2. By this action, *the charge packet, or "bucket,"* can be "handed down the line" (clocked) and eventually will cause a small negative spike in V_{out} at the detector (which is located at the end of the brigade). There are many variations on this theme, but these are the basic ideas involved in bucket-brigade devices.

Charge-Coupled Devices (CCD)

The operation of charge-coupled devices is similar, but the intermediate source-drain regions are omitted and minority-charge packets are simply transported along the surface of the semiconductor, as shown in Figure 7-37. The source-drain regions of the previously described BBD basically allow charge-transfer operation with relatively widespaced gate metal.

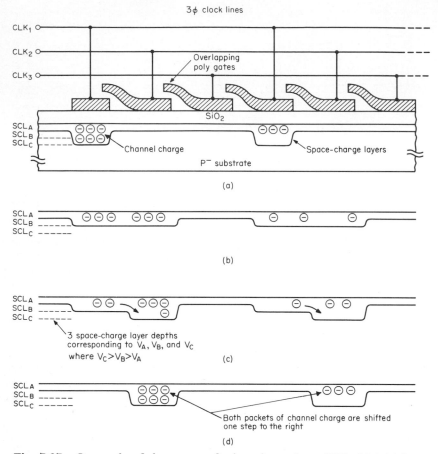

Fig. 7-37. One cycle of charge transfer in a three-phase CCD: (*a*) initial state ($t = 0$): $V_{CLK(1)} = V_C$, $V_{CLK(2)} = V_{CLK(3)} = V_A$; (*b*) at $t = t_1$: $V_{CLK(1)} = V_{CLK(2)} = V_B$, $V_{CLK(3)} = V_A$; (*c*) at $t = t_2$: $V_{CLK(1)} = V_B$, $V_{CLK(2)} = V_C$, $V_{CLK(3)} = V_A$; (*d*) at $t = t_3$: $V_{CLK(1)} = V_{CLK(3)} = V_A$, $V_{CLK(2)} = V_C$.

When polysilicon gates are available, direct gate overlap is possible. For simplicity, we have neglected to show how the initial mobile charge packets were introduced in Figure 7-37*a* (shown, for clarity, simply as 6 and 3 mobile electrons). These could have been injected electrically, as in the BBD example, or they could have resulted from electron-hole pair generation via photon bombardment, such as takes place in the optical imaging of charge-coupled imagers (CCIs). (CCI applications require a relatively long time in a static mode to allow optical integration of photon-induced channel-charge formation. The image-charge information is then relatively rapidly shifted out to avoid smearing the image.)

The CCD example in Figure 7-37 uses three different gate-drive voltage levels, and we have shown the resulting depletion or space-charge layers in an exaggerated scale for clarity. The different gate voltages are clocked to provide the time sequence of relative SCL formations shown. The first phase, at time t_1, redistributes any available mobile charge between the source cell and the next cell down the string. The gate voltage on the next cell is then increased to favor charge transfer during time t_2. When the gate voltages are switched and no channel charge is available, a wider SCL results than would exist if a channel were to form. This has been called *deep depletion* and occurs when the charges on the gate, which otherwise would terminate on channel charge, now reach deeper into the bulk to find the balancing fixed charges. Finally, one cycle is completed when the gate voltage of the original cell is reduced and the gate voltage of the adjacent cell is increased to the maximum value. Again, many derivatives of this basic concept exist; this discussion just presents the ideas involved.

7.8 MOSFET DYNAMIC RANDOM-ACCESS MEMORY (RAM)

Until 1972, main-computer memory was built with small magnetic cores because the cost of semiconductor memory was too high. The 1103 1-kb (1024 bits) dynamic MOS RAM (read-write, or random-access memory) was the turning point, since the cost and access time of this product was almost competitive with core. In MOS RAM, the bit capability has quadrupled every 4 years since 1972. The present standard is the 16-kb RAM; a 64-kb RAM is expected in volume production in 1981, and 256-kb in 1986. The highest bit densities and the most innovative design strategies can be found in the dynamic RAM products. Static RAM remains about one generation in bit density behind dynamic RAM.

Simplifying the RAM Cell

Early dynamic memory used only three transistors per bit, which resulted in a 4 : 1 increase in memory density over the more complex static-memory products. Many 1-kb RAMs used this cell. Next came the single transistor plus a storage capacitor; many 4-kb designs used this, and the cell size decreased to about 1 to 2 mil². The 16-kb memories initially used a one-transistor-per-cell design that was a byproduct of the charge-coupled devices and has therefore been called the *charge-coupled RAM cell.*

The basic 4-kb dynamic-memory cell, which uses a single transistor and a storage capacitor, is shown in Figure 7-38. It is really more of

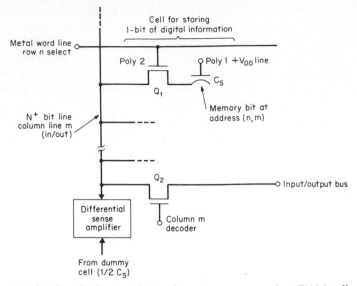

Fig. 7-38. Single transistor plus a storage-capacitor RAM cell.

an *analog sample-and-hold circuit* than a digital circuit. Since the digital information is held as a charge on a small-valued capacitor, the voltage must be restored every 2 ms because of the charge loss (or gain) from leakage currents. This is called *refreshing* the memory and is achieved by the use of a sense amplifier, which first compares the stored signal with a reference voltage to determine the stored-logic data. For example, if the signal voltage read from the storage cell exceeds the reference, the sense amplifier then regeneratively drives the storage capacitor back to a full V_{DD} or digital logic 1 level. If the signal from the stored cell is below the reference, the sense amplifier drives the storage capacitor to a 0-V level. Every time a row is selected, all the storage capacitors on that row are refreshed. Each capacitor in the memory array requires a refresh cycle at least every 2 ms, and this is why this device is called *dynamic memory.*

To write data into a cell, the particular row and column of the cell are first selected and the input/output (I/O) bus pin is then used to input the new digital data to the cell. For a READ operation, the same selection is made, but this time a voltage level is obtained from the cell as an output to the sense amplifier; this provides the logic signal for an output buffer which is then used to drive the I/O bus.

The layout of an early 4-kb dynamic RAM is shown in Figure 7-39. The storage capacitor is formed by the channel region of an MOS transistor. The column lines are N^+ diffusions. The output-signal voltage

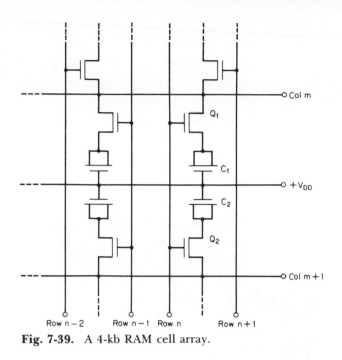

Fig. 7-39. A 4-kb RAM cell array.

from a cell can be increased by minimizing the parasitic capacitance of the column lines. This is accomplished by minimizing the row-to-row spacing (called the *row pitch*) and width of the N^+ strip. The column-line capacitance can be further reduced by preventing the higher concentration of the field dopant from diffusing into the N^+ region.

16-kb Dynamic RAM

The highly compact memory cell of the early 16-kb dynamic RAM made use of a double-poly silicon-gate process. Researchers C. N. Ahlquist, J. R. Breivogel, J. T. Koo, J. L. McCollum, W. G. Oldham, and A. L. Renninger received the 1976 International Solid State Circuits Conference Best Paper Award for their paper entitled "A 16384-Bit Dynamic RAM." Such a dynamic RAM made use of charge-coupled device (CCD) technology. If you remember the discussion of charge-transfer devices, you will recall that the CCD eliminated the extra diffusions in the bucket-brigade example. The same trick is used to expand from 4 kb, as shown in Figure 7-40, to 16 kb, as shown in Figure 7-41. The figures show only the details of the channel charge in simplified form. During a WRITE cycle, the poly-2 gate is taken positive. If the voltage applied to the source diffusion (the digital input) is at V_{DD}, no channel forms;

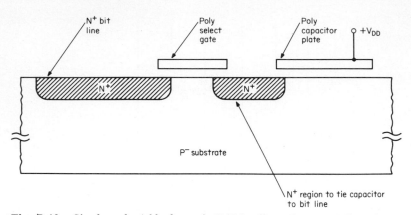

Fig. 7-40. Single-poly 4-kb dynamic RAM cell needs extra N⁺ region.

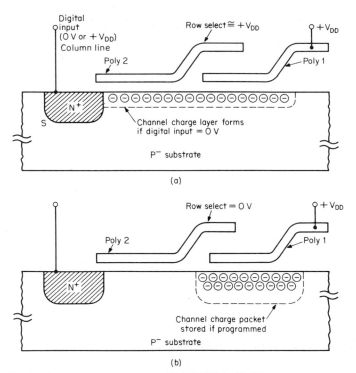

Fig. 7-41. The 16-kb charge-coupled RAM cell: (*a*) programming; (*b*) storing.

but if it is at 0 V, mobile electrons rush out of the source to form a channel. If a channel is formed, the channel packet is isolated under the poly-1 layer when the ROW SELECT gate (poly-2) goes to 0 V. So here is the digital information—and there is only 1/2 of a MOS device present.

To read the cell, the column lines are first precharged to V_{DD}. When the storage cell is subsequently gated onto this column line, an empty cell (indicating that a logical 1 was stored) will cause only a slight voltage loss. A cell with a charge packet present (indicating that a logical 0 was stored) will cause a larger drop in the precharged voltage level. It is the job of the sense amplifier to resolve this small voltage difference and to properly provide the correct digital output information.

The slightly strange thing is that the *refreshing operation keeps removing the channel-charge packet* that would otherwise form, given enough time. Such a dynamic RAM is therefore using the basic idea of the CCD memory and has essentially stopped interest in CCDs for memory applications.

As a design goal, the 16-pin package of the previous 4-kb RAM was to be used again for the 16-kb product to conserve printed-circuit-board (PC-board) space. The package pinouts are shown in Figure 7-42. The package will only accept a die width of 150 mils maximum, so the 16-kb chip was designed as 145×234 mils.

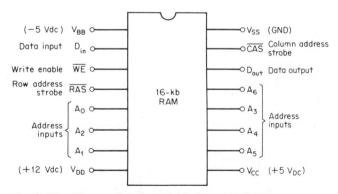

Fig. 7-42. Pinouts for the 16-kb dynamic RAM.

The key to small chip size is the layout of the RAM cell, since more than 16,384 of these will be used. The cells are arranged such that the N^+ region serves as the bit lines (or column lines) and the metal stripes, which contact the ROW SELECT gates (poly-2), are the row lines, as shown in Figure 7-43. Notice that two SELECT gates are picked up by only one metal-to-poly-2 contact.

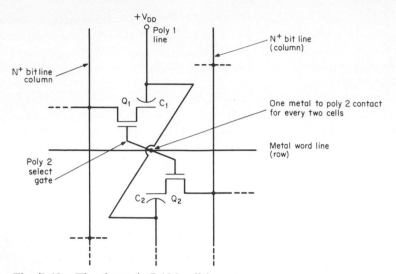

Fig. 7-43.   The dynamic RAM cell layout.

A detail of the cell layout (about 0.7 mil²) is given in Figure 7-44. This follows Figure 7-43 and shows how the two polysilicon layers, the N^+ bit lines, and the metal word lines (SELECT gates) are located on the chip. The poly-1 layer is permanently tied to V_{DD} and is used as the holding gate for the CCD stored-charge packet.

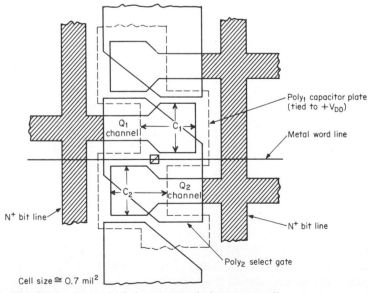

Fig. 7-44.   Layout of charge-coupled storage cell.

Dynamic RAM products, which use cells such as the ones in Figures 7-43 and 7-44, have a tendency to be susceptible to errors of a certain type. This effect is called *pattern sensitivity* and results from processing flaws that allow data-storage cells to be influenced by certain neighbors. For example, the cells in Figure 7-44 are more likely to pass channel charge between two adjacent cells C_1 and C_2 (which are shown) than between neighboring cells from adjacent columns (not shown) because the N^+ bit lines between the columns act as isolation barriers and do not allow charge coupling.

Sensing the Stored Bit

A high-speed differential sense amplifier is used to sense the stored bit and can be realized by using the input signal from the storage cell to slightly unbalance a flip-flop prior to power up. This circuit will regenerate the initial imbalance and produce a full logic swing at the output terminals. Such a circuit, shown in Figure 7-45, uses the cross-coupled transistors Q_7 and Q_8 to form the flip-flop gain devices. Transistors Q_1 and Q_2 are clocked (gated) loads, and transistor Q_9 serves as an ON, OFF control for the gain devices of the flip-flop. The storage cell C_S and a 1/2-value reference cell C_R are coupled to the flip-flop via transistors Q_3 and Q_5, respectively. A shorting transistor Q_4 ensures that the flip-flop starts out in a balanced condition.

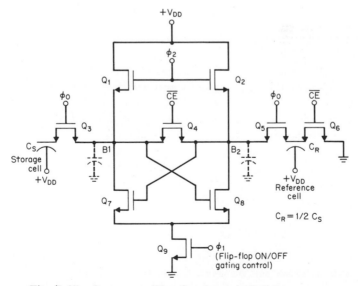

Fig. 7-45. Sense amplifier for dynamic RAM.

The timing of clocks ϕ_0 to ϕ_2 is referenced to the CHIP ENABLE signal $\overline{CE}$, which is an internally generated signal that is only slightly delayed from the external row-address strobe (RAS) used to initiate a RAM access. The circuit details of the timing sequence (the READ cycle of the RAM) are called out on the timing diagram in Figure 7-46. The

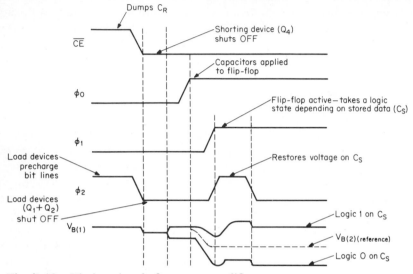

Fig. 7-46.   Timing signals for sense amplifier.

idea is to discharge the reference cell C_R at the start. With the flip-flop OFF (Q_9 not conducting), load transistors Q_1 and Q_2 are turned ON, while the shorting transistor Q_4 is also ON. This will precharge the parasitic capacitances of both bit lines B_1 and B_2 to the same voltage ($V_{DD} - V_{TH}$). Following this precharge, the capacitors are both simultaneously connected to the bit lines and a differential voltage will result between B_1 and B_2. This causes the flip-flop to take a logic state when it is subsequently turned ON via Q_9. Note that a little extra positive pulse on the ϕ_2 clock line is used to momentarily turn ON the load transistors so that the voltage of the storage capacitor C_S can be restored. This is the way the RAM is refreshed. As long as the voltage on C_S can be properly read, it will be properly refreshed to the correct logic value.

α-Particles and Soft Errors

The year 1978 will be remembered as the start of *α-particle worry*. Both static and dynamic RAMS were found to be susceptible to *soft failures*, or the spontaneous flipping of a memory bit, while the basic IC is operating properly. This is in contrast to *hard failures,* where the IC is made inoperative by the presence of stronger radiation. On further investigation of the soft-failure mechanism, an *upset charge* (the amount of electrons necessary to cause a memory bit to fail) was found to be generated within the silicon as a result of α particles.

These α particles (helium nuclei) can be provided by the action of cosmic rays or can come directly from the materials of the IC package. A moving charge is necessary to cause problems in silicon, and the ionization wake of these particles leaves behind a trail of electron-hole pairs within the crystal. In addition, the particles can directly interact with silicon nuclei and cause other charged particles to be generated, and this also creates a dense burst of electron-hole pairs.

As the size of the circuits is reduced, the upset charge is also reduced, since there are now fewer electrons stored in the circuit. Therefore, sensitivity to α particles is of major concern in the emerging VLSI circuit designs and process technologies.

There are no α particles in cosmic rays. The α particles are produced as a secondary reaction within the silicon and are due to the generation of elementary particles by the cosmic rays. These particles, such as neutrons, protons, pions, electrons, and muons (heavy electrons with the charge of an electron but 200 times more massive), cause soft errors in memory products. The problem is not yet properly modeled, and much work remains before accurate predictions can be made.

The source of the α particles, whether from the packaging materials or from cosmic rays, can be determined if an IC with a soft-error problem is (1) taken up a mountain or in an airplane to increase cosmic rays (failure rates are up by a factor 4 or 5 in Denver, Colorado, for example), (2) sunk in a mine or water tank to reduce cosmic rays, or (3) placed inside a 10-in concrete shield (which, strangely enough, causes cosmic-ray-induced error rates to increase).

Other measures now taken to reduce this problem are (1) increase the memory-storage capacitor so it can accommodate the electrons lost or gained by one α particle (10^6 electrons) and (2) cover the chip surface with a silicone-protecting material.

It is interesting that the collector-base SCL of the parasitic vertical bipolar NPN transistor of the CMOS process can collect the generated electrons, passing them to the $+V_{CC}$ supply, and thereby protect the critical N^+ source-drain memory-storage nodes. In NMOS, the generated electrons add to the channel charge and tend to cause higher soft-error failure rates.

256-kb Dynamic RAM

As an example of VLSI, a 256-kb single-transistor-cell RAM in a standard 16-pin package has been reported. This RAM uses 1.5-μm lithography (an optical direct wafer stepper) and makes use of dry-etching techniques. The 5.7- $\times$ 12.5-μm memory cell uses a 200-Å oxide thickness for the storage cell (to increase the capacitance in order to prevent a soft error owing to only one α particle) and 400 Å for the transistor gate region, as shown in Figure 7-47. A data-access time of 160 ns ensures compatibility with the latest microprocessors, and the 4.84- $\times$ 8.59-mm (191- $\times$ 338-mil) chip consumes only 45 mA at 5 V when active and 5 mA during standby.

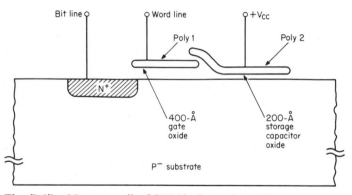

Fig. 7-47.   Memory cell of 256-kb dynamic RAM.

7.9 STATIC NMOS RAM

Designers of static RAMs are also using the new process technologies. To obtain a stable memory cell that does not need to be refreshed, a flip-flop circuit is used for each bit of a static memory, as shown in Figure 7-48. This 6-transistor circuit uses extremely small-sized, high impedance load transistors Q_1 and Q_2 to minimize the power drain.

All flip-flop storage circuits require accessing both sides with complementary bit lines. For example, if the bit line on the left side is at 0 V, the right-side bit line will be at 5 V. When word-line coupling transistors Q_3 and Q_4 are pulsed ON, the drain of Q_5 will be pulled low and the drain of Q_6 will be forced high. By driving both sides, we can rapidly write the cell with new data.

Newer static RAMs have reduced this cell size and thereby have achieved 64-kb memories by replacing the load transistors with resistors provided by a second polysilicon layer that is nearly intrinsic (only lightly doped). These polysilicon load resistors change drastically with tempera-

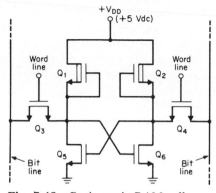

Fig. 7-48. Basic static RAM cell.

ture (2 to 3 orders of magnitude) and typically have a minimum value of 10 MΩ. This not only reduces the power drain, but also allows these load resistors to be located in the vertical dimension of the die, which results in a smaller cell size.

As an example of future products, a 64-kb static RAM has been reported that also makes use of poly-R's in the standard flip-flop storage cell, which has been kept small (16 × 19 μm) by using 2-μm design rules. The completed chip (5.44 × 5.80 mm, or 214 × 228 mils) draws 60 mA (which drops to 15 mA during standby) from a 5-V supply and has an access time of 80 ns. For even lower power during standby, a bulk CMOS 16-kb static RAM also has been reported that draws 40 mA at 5 V when active and only 1 μW during standby. This 5.06- × 5.77-mm (199- × 227-mil) IC memory has an access time of 35 ns.

7.10 NONVOLATILE SEMICONDUCTOR MEMORY

The first nonvolatile semiconductor memory was the mask-programmable read-only memory (ROM). Early applications of these ROMs were mostly for look-up tables. Use has since expanded, because ROM is the popular μP program-storage medium. Because this code typically changes both during project development and during early production runs, a need was created for a reprogrammable ROM. This need was largely supplied by the ultraviolet (UV) erasable EPROM circuits.

It has long been known that a layer of silicon dioxide (SiO_2) will not stop ionic contamination (such as sodium) and that silicon nitride (Si_3N_4) will. It was found that if this nitride is placed directly on a high-resistivity (lightly doped) region of silicon, excessive leakage currents will result. To prevent this, one early PMOS researcher first put down a thin layer (50 Å) of SiO_2, which was then followed by a thicker layer (500 Å) of Si_3N_4 to form the gate dielectric. This sandwich structure was supposed to prevent gate-region ionic contamination. Unfortunately, another phenomenon resulted: the gate dielectric region would trap charge (electrons), and this would result in a large shift in threshold voltage and cause the PMOS transistor to be always ON. In experiments with this V_{TH} shift, it became obvious that the gate dielectric could store charge for an extremely long time. Thus the first nonvolatile memory was born, and this was the starting point for EPROM circuits.

UV Erasable Programmable ROMs (EPROMs)

With silicon-gate transistors it was discovered that if the drain junction was driven to avalanche breakdown, high-energy electrons (*hot electrons*) would penetrate a relatively thick gate-oxide layer and become trapped on the gate. The floating gate could then hold onto the trapped electrons for extremely long intervals of time (thousands of hours to years). The first devices of this type were called *floating-gate avalanche-injection MOS transistors* (FAMOS), and they were built using a single polysilicon process and needed an additional transistor to read the state of the floating-gate storage cell.

A 2-kb FAMOS P-channel EPROM was introduced in 1972. The erase is provided by shining ultraviolet light onto the chip through a clear quartz lid. This excites the electrons and causes them to move off the gate.

Special double-poly N-channel silicon-gate storage transistors were next made available (with 8-kb capacity), and these used a floating first-layer polysilicon gate and could be directly accessed via a second stacked polysilicon gate, as shown in Figure 7-49. Testing the memory cell amounts to applying a special-amplitude sense voltage to the second polysilicon layer or control gate. If the cell was programmed, the first-layer polysilicon gate will cause a drastic shift in the threshold voltage, as shown in Figure 7-50. Drain current *will flow* if the gate *was not programmed* or will *not flow* if the cell *was programmed* (the electrons on the gate keep the N-channel transistor OFF). These double-poly circuits are called *stacked-gate EPROMS*. Further work in this area added the enhancement-and-depletion mode (E/D) NMOS processes, and these became very popular because they were made available at 16- and 32-kb capacities.

The continuing need for EPROM products in the new generation μP's has stimulated interest, and a 64-kb product has been reported

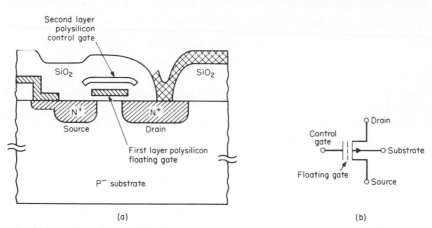

Fig. 7-49.   Two-layer polysilicon stacked-gate EPROM: *(a)* cross section; *(b)* symbol.

(128 kb is close, and 256 kb is being designed). The 64-kb chip makes use of scaling in the N-channel EPROM process to reduce the cell area to 0.25 mil² and has an access time of less than 200 ns (less than 1/2 the time required by nonscaled EPROMS). This 8-kb $\times$ 8-b memory chip is 179 $\times$ 181 mils and is in a 28-pin package. Active power is less than 500 mW and standby is less than 100 mW.

Unfortunately, the only way to remove the trapped *gate electrons* from all these EPROMS is by irradiation through a relatively high-cost clear quartz window (on a ceramic package) with relatively high-intensity ultra-violet light. New developments in this area are electrically erasable PROMs (called EEPROMS, or E²PROMs); they will be considered next.

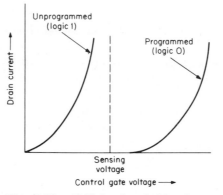

Fig. 7-50.   Shift in threshold voltage of an EPROM.

Electrically Erasable PROMS (E²PROMs)

There are two types of floating-gate E²PROMs, and the IC industry is not far down on either learning curve. One makes use of reversible electron tunneling through a thin oxide layer and the other relies on sharp projections (called *asperites*) between triple polysilicon layers to allow electron tunneling both onto and off of a floating gate. Both these E²PROM technologies will be tested in the 1980s. We will consider the reversible tunneling E²PROM first.

16-kb Tunneling E²PROM   A 16-kb E²PROM has been reported that makes use of reversible electron tunneling through a thin SiO_2 layer to both program and erase a floating polysilicon gate, as shown in Figure 7-51. This E²PROM uses voltages less than 25 V, and the low current

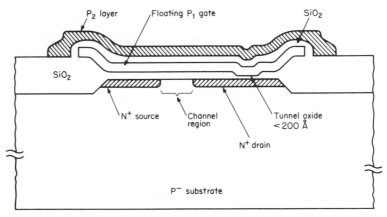

Fig. 7-51.   Tunneling E²PROM cell.

drain allows future products to provide this high voltage by capacitive "bootstrap" circuits located directly on the IC chip.

The floating gate is charged by providing a 20-V signal to the P_2 layer (75 % of this voltage capacitively couples to the P_1 floating gate) while the drain is held at ground. To discharge the floating gate, the leads are simply reversed: 20 V to the drain and ground to gate P_2. Another basic E²PROM involves programming by passing electrons between a triple polysilicon stack. This will be discussed next.

Asperitic E²PROM   We have discussed how electrons can tunnel through thin (50-Å) gate oxides. If 1000-Å SiO_2 layers are used between multiple polysilicon layers that are specially processed to have small sharp projections (called *asperities*) on their surfaces, sufficiently high electric fields for tunneling can be obtained at less than 25 V. On-chip capacitive voltage-stepup circuits are used to supply this high voltage, because only small currents are required. Further, these 1-kb memories operate from a single 5-V power supply.

Three polysilicon layers are used: the first provides a ground reference plate, the second becomes the *floating gate,* and the third is the PROGRAM/ERASE control line. Programming the floating gate (Figure 7-52) is accomplished by applying the high voltage to the third polysilicon

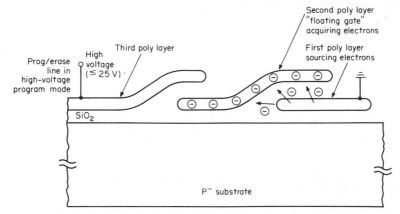

Fig. 7-52.   Programming an E²PROM cell. Programming the poly-2 layer does not require a FET; it is accomplished on the surface of the chip.

control layer. Capacitive coupling to the second layer causes this floating gate to acquire electrons from the first layer. When the programming line is returned to 0 V, the floating gate assumes a negative voltage because of the presence of these trapped electrons. This voltage is retained, even if the IC power supply is turned OFF.

The E²PROM cell is erased as shown in Figure 7-53. A transistor

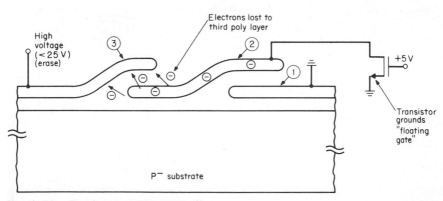

Fig. 7-53.   Erasing an E²PROM cell.

is used to ground the floating gate and the PROGRAM/ERASE line is again taken high. Electrons now leave the floating gate and travel to the PROGRAM/ERASE line. When this control line is returned to 0 V, the floating gate will also be at 0 V.

The combination of the floating-gate transistor and the control elements is shown in Figure 7-54. This strange structure is used to control

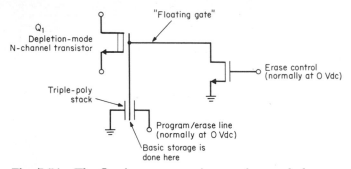

Fig. 7-54. The floating-gate transistor and control elements of the E²PROM.

the conduction of depletion-mode transistor Q_1. If the floating gate is at a negative voltage level, Q_1 is held OFF. Transistor Q_1 will be ON if the gate is at 0 V.

This PROM structure is embedded within a static RAM circuit, as shown in Figure 7-55. A standard RAM is modified by adding the float-

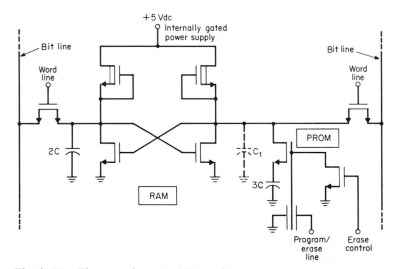

Fig. 7-55. The complete E²PROM cell.

ing-gate transistor to only one side of the storage flip-flop. The weighted capacitor values shown are the key to the operation of the circuit. The capacitor C is the stray capacitance that unavoidably exists at the flip-flop output node.

The PROM logic level stored is determined by the potential of the floating-gate electrode. If this gate is at the negative voltage, the associated transistor is held OFF and the $3C$ capacitor (which is 3 times larger than the stray capacitance C) is isolated from the flip-flop. Now, when the internal 5-V supply to the RAM cell is cycled to 0 V and back to 5 V, the cell will *wake up* with the left side of the flip-flop at a low voltage level because of the larger delaying capacitance at this node ($2C$ versus C). Conversely, if the floating gate is at 0 V, the larger $3C$ capacitor now loads the right side of the flip-flop, which causes the reverse-logic state. In this manner, the PROM data can be loaded into the RAM at any time and is automatically loaded when the IC is first powered up. Further, the contents of the RAM can be written into the PROM at any time. As can be seen, this is a very versatile memory component.

7.11 A LOOK INTO THE FUTURE

As we look into the future, we can expect that process refinements will still be vigorously pursued. Improvements in mask making and lithography are expected. Size is the key to increasing speed and allowing more function in a given die area to ensure lowest cost. Channel lengths have been reduced to less than 2 μm with gate-oxide thicknesses of 400 Å. Even smaller devices have been reported (channel lengths of 0.8 μm) and will soon be in production.

Technologists have indicated that a limit exists to how thin the gate oxide can be made; for example, 200 Å will not allow operation at the current 5-V power-supply level. This will therefore force a reduction to about the 2-V level in the near future to allow additional device scaling. (The optimum supply voltage has been estimated to be 0.5 V_{DC} to take *full* advantage of device scaling.)

The production-lithography standard is the optical-projection scan machine. These machines can achieve less than 2-μm line widths. Automatic pattern-recognition and edge-detection electronics are presently permitting alignment accuracies of 0.1 μm.

Improvements are possible by using optical-projection stepping machines, which are expected to allow 0.5-μm line widths in the future. These machines increase the cost of an alignment and present economic problems in their widespread use.

The highest resolution is obtained with an E-beam system that writes directly on the wafers. This can achieve submicrometer resolution but is considerably slower and costs very much more than other devices. Again, there are economic and machine-availability problems, but these machines will be useful for expediting the first wafers out of the fabrication line without the delays of mask making.

To determine the limits on size and power dissipation, researchers have asked fundamental questions based on the physics of information processing by considering the basic theoretical structures of thermodynamics and information theory. The main question is: What energy dissipation is required to achieve information processing? These studies have indicated that even the most pessimistic estimates show that modern circuits are many orders of magnitude higher in dissipation than the fundamental theories indicate.

The success of the MOS processes is expected to affect the future of the traditionally bipolar linear IC circuits. The question now is how to make use of the strengths of MOS in linear designs. *Multiple low-cost switches are the new thing,* and they are creating a strong interest in sampled-data systems. Many linear functions are possible with a sampled-data approach, and many performance advantages are obtained as compared with existing linear circuits. We are already seeing switched-capacitor filters, sampled-data comparators, sampled-data amplifiers, and even *dynamic amplifiers.*

High-power discrete FET devices are also available with performance advantages over bipolar-power transistors, and MOS-power ICs have not yet been fully exploited. MOS devices have a simpler breakdown-voltage mechanism (no second breakdown exists), and the current of the MOS device reduces as the junction temperature increases. It appears that we are leaving the era of the bipolar transistor and entering the new *MOS age of electronics.*

Bibliography

1. Adler, R. B., A. C. Smith, and R. L. Longini: *Introduction to Semiconductor Physics,* vol. 1., Semiconductor Electronics Education Committee, Wiley, New York, 1964.
   A recommended text to provide additional information and a more rigorous development of the material without losing the intuitive approach. It covers much of solid-state physics in a way that an electronics engineer can follow.

2. Calder, N.: *The Key to the Universe—A Report on the New Physics,* Viking Press, New York, 1977.
   This version of a British Broadcasting Company report on the recent findings of high-energy physicists is an excellent presentation of the new theories on atomic structure that were advanced during the 1970s and later confirmed by particle-accelerator experiments. A fast-moving story of the particles, forces, theories, people, and machinery behind this fast-paced science. It is useful reading to become more familiar with the electrons, protons, and neutrons of the physical world.

3. Carroll, J. E.: *Physical Models for Semiconductor Devices,* Crane, Russak, New York, 1974.
   This book is recommended because it carries on an intuitive development similar to the one in this book and adds more of the physics and mathematics.

4. Gray, P. R., and R. G. Meyer: *Analysis and Design of Analog Integrated Circuits,* Wiley, New York, 1977.
   Recommended for those who are interested in circuits and applications of analog (or linear) ICs.

5. Gamov, G.: *The Atom and Its Nucleus,* Prentice-Hall, Englewood Cliffs, New Jersey, 1961.
   A very readable and interesting discussion of the history and significant breakthroughs of atomic physics. It is highly recommended as a first book for those who want to pursue the physics of the solid state.

6. Gamov, G.: *Thirty Years That Shook Physics,* Doubleday, Garden City, New York, 1966.
   A very interesting story of the people who brought in quantum mechanics. It is highly recommended to all readers and puts the ideas and problems of the new physics in perspective. It provides good background material for those who want to increase their knowledge of physics.

7. Ghandhi, S. K.: *Semiconductor Power Devices,* Wiley-Interscience, New York, 1977.
   One of the few books devoted to semiconductor power devices. It brings up the contrasts

between semiconductor devices and monolithic ICs. In addition, it covers new significant devices in detail and is recommended for anyone who uses or designs high-power semiconductor devices.

8. Gray, P. E., D. DeWitt, A. R. Broothroyd, and J. F. Gibbons: *Physical Electronics and Circuit Models of Transistors,* vol. 2., Semiconductor Electronics Education Committee, Wiley, New York, 1964.

An excellent text on PN junctions and bipolar transistors. It covers the physics, limitations, and modeling of these structures. It is highly recommended to people who work with these components and is especially valuable to circuit designers.

9. Grove, A. S.: *Physics and Technology of Semiconductor Devices,* Wiley, New York, 1967.

One of the first of the early texts that concentrated on silicon as opposed to germanium. Still a classic reference text, it is easy to read and has good discussions of JFET and MOSFET devices as well as much of the basic physics of solid-state devices.

10. Hutchison, T. S., and D. C. Baird: *The Physics of Engineering Solids,* 2d ed., Wiley, New York, 1968.

A good text on the fundamental properties of the solid state. It provides both a good word description and the supporting mathematics.

11. Muller, R. S., and T. I. Kanis: *Device Electronics for Integrated Circuits,* Wiley, New York, 1977.

One of the better modern texts on solid-state devices. It has good descriptions and discussions of energy bands and metal-semiconductor contacts, as well as bipolar and MOS devices. It also includes a wide range of useful material on processing and modern semiconductor processes.

12. Yang, E. S.: *Fundamentals of Semiconductor Devices,* McGraw-Hill, New York, 1978.

This book has a very wide range of information, from the basics to the most current semiconductor devices. The chapters on metal-semiconductor junctions, solar cells, light-emitting diodes, and charge-transfer devices are especially good. The appendixes on atoms, electrons, and energy bands are also well done. This is the best single reference text and the explanations have benefited from the latest theories of semiconductor physics.

Index

About the Author

Thomas M. Frederiksen is a linear IC design engineer with the National Semiconductor Corporation, Santa Clara, California. Upon earning his BSEE degree from California State Polytechnic University at San Luis Obispo, he started his professional career as a development engineer with the Motorola Systems Development Laboratory. Subsequently he worked with the Microelectronics Group at Hughes Semiconductor Division and later became senior project engineer at Motorola Semiconductor Products Division.

In 1971 Mr. Frederiksen joined his present firm where he developed custom ICs and standard single-supply building block circuits for automotive and industrial applications. Currently he is involved with data acquisition circuits that will interface with microprocessors and is designing a standard line of A/D converters.

Mr. Frederiksen holds more than 36 patents on linear ICs and devices, is a frequent contributor to professional literature, and has given a number of seminars on the subject in the U.S. and abroad. In 1977 he received the International Solid State Circuits Conference Best Paper Award.